Astrophysics and Space Science Proceedings

International Heliophysical Year
IHY 2007
Developing Nations Participation in International Global Studies of the Sun-Earth System through the United Nations
http://ihy2007.org

Proceedings of the Third UN/ESA/NASA Workshop on the International Heliophysical Year 2007 and Basic Space Science

National Astronomical Observatory of Japan

H.J. Haubold
Editor

United Nations Office for Outer Space Affairs (UNOOSA), Austria

A.M. Mathai
Editor

McGill University Dept. Mathematics & Statistics, Canada

Springer

Editors
Prof. Dr. H.J. Haubold
United Nations Office for Outer Space Affairs (UNOOSA)
Vienna International Centre
Wagramer Str. 5
1400 Wien
Austria
hans.haubold@unoosa.org

Prof. Dr. A.M. Mathai
McGill University
Dept. Mathematics & Statistics
805 Sherbrooke St. W.
Montreal QC H3A 2K6
Canada
mathai@math.mcgill.ca

ISSN 1570-6591 e-ISSN 1570-6605
ISBN 978-3-642-03323-0 e-ISBN 978-3-642-03325-4
DOI 10.1007/978-3-642-03325-4
Springer Heidelberg Dordrecht London New York

Library of Congress Control Number: 2009933994

Cover design: eStudio Calamar S.L.

Printed on acid-free paper

Springer is part of Springer Science+Business Media (www.springer.com)

Preface to the Proceedings of the UN/ESA/NASA Workshop

From 6 to 15 June 2007, the Committee of the Peaceful Uses of Outer Space (COPUOS) held its fiftieth session in Vienna, Austria, marking the important convergence of anniversaries, including the 50th anniversary of the space age, the 40th anniversary of the enforcement of the Outer Space Treaty and the celebration of International Heliophysical Year 2007 (IHY 2007). COPUOS, which is the only permanent UN body that deals exclusively with peaceful uses of outer space, was established in 1958 as an ad hoc committee by the UN General Assembly resolution 1348, which noted the success of the scientific co-operative programme of the International Geophysical Year (IGY 1957) in the exploration of outer space and decided to continue and expand this type of cooperation. In 1959, the Committee became a permanent body with a mandate to review the scope of international cooperation in the peaceful uses of outer space, devise programmes in this field to be undertaken under UN auspices, encourage continued research and the dissemination of information on outer space matters and to study legal problems arising from the exploration of outer space (http://www.unoosa.org/oosa/events/copuos50.html).

The UN General Assembly, in its resolution 60/99 of 2005, noted with satisfaction the contribution being made by the Scientific and Technical Subcommittee of COPUOS and the efforts of Member States and the Office for Outer Space Affairs (OOSA) to promote and support the activities being organized within the framework of IHY 2007 in the spirit of IGY 1957. IHY 2007 is now taken as an opportunity to (a) advance the understanding of the fundamental heliophysical processes that govern the Sun, Earth, and heliosphere, (b) continue the tradition of international research and advancing the legacy of IGY 1957, and (c) demonstrate the beauty, relevance and significance of space and earth science to the world (http://ihy2007.org, see also J. Davila et al. Space Policy 23(2007)121–126).

In preparation of IHY 2007, the OOSA (http://www.unoosa.org/oosa/en/SAP/bss/ihy2007/index.html), in cooperation with NASA, ESA, and the IHY Secretariat, hold international workshops in the United Arab Emirates in 2005 (http://www.ihy.uaeu.ac.ae/), in India in 2006 (http://www.iiap.res.in/ihy/), and at the National Astronomical Observatory of Japan in Tokyo, 18–22 June 2007 (http://solarwww.mtk.nao.ac.jp/ UNBSS_Tokyo07/). This workshop was a welcome opportunity to review achievements of this sort of workshops, organized annually by the United Nations since 1991.

The starting date of IHY 2007 was February 19, 2007. On that date, during the session of the Scientific and Technical Subcommittee of COPUOS, the IHY kick-off included an IHY exhibit, press briefing, and an opening ceremony in the United Nations Office in Vienna (http://ihy2007.org/newsroom/opening_ceremony.shtml). IHY regional coordinators, Steering Committee members and Advisory Committee members participated in the IHY kick-off event. The Austrian Academy of Sciences hosted a one-day symposium on IHY 2007 in Vienna on 20 February, 2007.

IGY 1957 was one of the most successful international science programmes of all time and broke new ground in the development of space science and technology. Fifty years later, IHY 2007 continues this tradition. The tradition of international science years began almost 125 years ago, when the first international scientific studies of global processes of the earth's poles took place from 1882 to 1883. A second International Polar Year was organized in 1932, but a worldwide economic depression curtailed many of the planned activities.

Many of the IHY 2007 participants hope for an amplification of their efforts when humanity will celebrate the International Year of Astronomy in 2009 (http://www.astronomy2009.org/).

The Proceedings of the UN/ESA/NASA Workshop, hosted by the National Astronomical Observatory of Japan, are published in two volumes. Volume I, containing 47 papers dealing with issues related to IHY, are brought out as a special issue of the international journal *Earth, Moon, and Planets*. The 47 papers also encompass those presented at the Second European General Assembly on IHY, held at Torino, Italy, 18–22 June 2007. Volume II, containing the 16 papers in this special issue of the international journal *Astrophysics and Space Science*, covers two programme topics pursued in this and past workshops of this nature: (a) non-extensive statistical mechanics as applicable to astrophysics and (b) the TRIPOD concept, developed for astronomical telescope facilities.

Equations governed by fractional calculus describe anomalous reaction and anomalous diffusion, thus bringing new mathematics and physics to well-known fundamental laws such as Debye relaxation and Fick's equilibration of concentration of species. It was Einstein who derived the diffusion equation from a probabilistic point of view central to Boltzmann–Gibbs extensive statistical mechanics. However, even Boltzmann–Gibbs statistical mechanics has been generalized, recently, to Tsallis nonextensive statistical mechanics, suitable for tackling astrophysical systems subject to spatial and/or temporal long-range interactions and memory. The first eleven papers in these Proceedings deal with nonextensivity of statistical mechanical systems and formalism for spatial-temporal phenomena as applicable to astrophysical plasmas. These papers address q-distribution, fractional reaction and diffusion, and reaction coefficient, as well as Mittag-Leffler function, all of which are important to the aforementioned phenomena. Subsequently, five papers address results of the TRIPOD concept for astronomical telescope facilities developed in workshops as this one since 1991.

Office for Outer Space Affairs, United Nations *H.J. Haubold*
Centre for Mathematical Sciences, India *A.M. Mathai*

Contents

T. Dauxois' "Non-Gaussian Distributions Under Scrutiny" Under Scrutiny

Constantino Tsallis

Abstract A recent paper by T. Dauxois entitled "Non-Gaussian distributions under scrutiny" is submitted to scrutiny. Several comments on its content are made, which constitute, at the same time, a brief state-of-the-art review of nonextensive statistical mechanics, a current generalization of the Boltzmann-Gibbs theory. Some inadvertences and misleading sentences are pointed out as well.

Keywords Nonadditive entropy · Nonextensive statistical mechanics · q-Gaussians · q-Central Limit Theorem

T. Dauxois has recently commented (Dauxois 2007) an interesting paper by Hilhorst and Schehr (Hilhorst and Schehr 2007). In the latter paper, two specific probabilistic models, namely a discrete one (Moyano et al. 2006) (from now on referred to as the *MTG model*) and a continuous one (Thistleton et al. 2006) (from now on referred to as the *TMNT model*), are *analytically* discussed. Both models consist in N correlated random variables (respectively discrete and continuous), and the point under study is what is the limiting distribution when $N \to \infty$. *Numerical indications*[1] have been found that quite strongly suggest that these limiting distributions could be of the q-Gaussian class ((1) in (Dauxois 2007)). Hilhorst and Schehr have shown (Hilhorst and Schehr 2007) that they are *not*, even if they are numerically intriguingly close. The relations between the entropic indices q (of the possible q-Gaussian limiting distributions) and the parameters of the models have been addressed as well. It was analytically confirmed in (Hilhorst and Schehr 2007) the correctness of the relations conjectured respectively in (Moyano et al. 2006) and in (Thistleton et al. 2006), if these *analytically exact* non q-Gaussian distributions were to be approached by q-Gaussians!

C. Tsallis (✉)
Centro Brasileiro de Pesquisas Fisicas and National Institute of Science and Technology for Complex Systems, Rua Xavier Sigaud 150, 22290-180 Rio de Janeiro-RJ, Brazil
and
Santa Fe Institute, 1399 Hyde Park Road, Santa Fe, NM 87501, USA
e-mail: tsallis@cbpf.br

[1] See the title of the present (Moyano et al. 2006).

H.J. Haubold and A.M. Mathai (eds.), *Proceedings of the Third UN/ESA/NASA Workshop on the International Heliophysical Year 2007 and Basic Space Science*, Astrophysics and Space Science Proceedings, DOI 10.1007/978-3-642-03325-4_1,

Why may all this have some general interest? The reason lies on the fact that q-Gaussian distributions play a special role in nonextensive statistical mechanics (Tsallis 1988; Curado and Tsallis 1991; Tsallis et al. 1998), a current generalization (based on the entropy (2) of (Dauxois 2007)) of the celebrated BG theory. Recent reviews of q-statistics and a vast set of experimental, observational and computational applications and/or verifications can be seen in (Gell-Mann and Tsallis 2004; Boon and Tsallis 2005; Abe et al. 2007; Tsallis 2009)[2]. Very specifically, it has been recently *proved* (Umarov et al. 2008) a q-generalization of the Central Limit Theorem for $q \geq 1$. More precisely, if we have N random variables that are strongly correlated in a special manner (called *q-independence*; see (Umarov et al. 2008) for details), it can be rigorously proved that their sum approaches, when appropriately centered and scaled, a q-Gaussian distribution. The standard CLT is recovered as the $q = 1$ particular instance. The proof for $q < 1$ has been provided in (Nelson and Umarov 2008). Since the *MTG* and *TMNT* models have a compact support (i.e., of the $q < 1$ type), one expects q-Gaussians whenever the correlation is of the q-independent type. Given the results in (Hilhorst and Schehr 2007), one is naturally led to argue that the strong correlations in those two models are *not q-independent but only almost q-independent.* An interesting question remains then open. What physical ingredient have the *MTG* and *TMNT* models failed to incorporate? In other words, there is something which is still missing in those models in order for the correlation to be *exactly q-independent.* What is it? Progress is presently being achieved along this line (deeply related to asymptotic scale-invariance; for models that have been analytically proved to approach q-Gaussians see A. Rodriguez, V. Schwammle and C. Tsallis, Strictly and asymptotically scale-invariant probabilistic models of N correlated binary random variables having q-Gaussians as N -> infinity limiting distributions, JSTAT P09006 (2008), and R. Hanel, S. Thurner and C. Tsallis, Limit distributions of scale-invariant probabilistic models of correlated random variables with the q-Gaussian as an explicit example, Eur. Phys. J. B (2009), DOI: 10.1140/epjb/e2009-00330), but this remains outside the present scope, which primarily is the careful scrutiny of the paper (Dauxois 2007). Let us address now some of the weaknesses of that paper.

(i) In contrast with what one reads in (Dauxois 2007), the *TMNT* model is *not* at all described nor presented in its (Moyano et al. 2006). In fact, it is not even mentioned there, and has never been published![3]. Dauxois visibly confuses with (Thistleton et al. 2007), whose content *has absolutely nothing to do* with the model discussed by Hilhorst and Schehr. Had the Author of (Dauxois 2007) paid more attention to his own (Moyano et al. 2006), or at least to its title, the confusion would have been avoided.

[2] A regularly updated bibliography is available at http://tsallis.cat.cbpf.br/biblio.htm

[3] The details of the *TMNT* model, as well as all our numerical and graphical results, were transmitted privately by me to Hilhorst, who had heard, in Natal-Brazil in March 2007, an oral presentation of mine including that subject

(ii) The statement "*could be the basis for a generalized central limit theorem*" (below (1) of (Dauxois 2007)) reveals that the Author is possibly unaware of the proof existing since already some time in (Umarov et al. 2008) (further extensions are possible along the lines of (Umarov et al. 2008; Umarov and Tsallis 2007)).

(iii) The statement "*The basis for this suggestion* ..." (above (2) of (Dauxois 2007)) reveals that the Author is unaware that the crucial point is *not* the isolated fact that q-Gaussians optimize S_q (something which, in contrast with what is referred in (Dauxois 2007), is *not even vaguely mentioned* in (Hilhorst and Schehr 2007) of (Dauxois 2007)), but rather the remarkable fact that q-Gaussians also happen to be (Plastino and Plastino 1995; Tsallis and Bukman 1996) *exact* stable solutions of the nonlinear Fokker-Planck equation (since many decades called *Porous Medium Equation*). The mere fact that q-Gaussians would optimize some specific entropic functional would certainly not be of any particular significance. The entire rationale of what was at the time a conjecture (and is now a proved theorem) can be seen in (Tsallis 2005).

(iv) The qualification "*in the absence of firm grounds...*" reveals that the Author of (Dauxois 2007) is possibly unaware of the numerous rigorous results which precisely provide a firm mathematical basis for the entropy S_q and the associated nonextensive statistical mechanics. Among many others, let us mention the q-generalization, respectively in (Santos 1997) and in (Abe 2000), of Shannon's and of Khinchine's uniqueness theorems, the already mentioned q-central limit theorems in (Umarov et al. 2008), the Lesche-stability (Abe 2002) and the Topsoe-factorizability (Topsoe 2006) of the entropy S_q, the analytical results in (Anteneodo and Tsallis 2003) for Langevin-like stochastic equations in the presence of multiplicative noise, the analytical results of (Baldovin and Robledo 2002a,b, 2004; Mayoral and Robledo 2005) concerning unimodal dissipative one-dimensional maps, the analytical connection (Wilk and Wlodarczyk 2000; Beck 2001; Beck and Cohen 2003) to the Beck-Cohen superstatistics, etc.

(v) The Author of (Dauxois 2007) wonders whether the whole idea of q-statistics could be "*just a nice idea and a powerful fitting function*". Fitting functions typically have one or more fitting parameters. Let us focus on this point. Fitting parameters can be of quite different natures. The first, and in some sense more fundamental, class is constituted by those numbers for which no theory exists in contemporary physics that could produce them from more basic principles. In this class we have all the so-called *universal constants* of contemporary physics (c, h, G, and k_B), which, as well known, might not even be constants! They are obtained, nowadays with great precision, from fittings concerning basic physical laws such as the Planck law for the black-body radiation. A second class is constituted by those parameters that can ideally be deduced from first principles through some specific way involving some specific information. But, in many cases, the procedure turns out to be untractable, either because of mathematical difficulties, or because some of the necessary information is missing. Such is the case of the orbit of say Mars.

Newtonian mechanics can in principle predict this orbit given, at some time, the positions and velocities of all masses of the planetary system. Since this information is unavailable (and even if it was available, a colossal computer would be needed to perform the calculations!), astronomers use, as a first approximation, the elliptic form of a Keplerian orbit and just fit this form (which can be easily obtained within Newtonian mechanics) in order to obtain the concrete values of the two axis of the ellipse. The entropic index q belongs to this class. It can be in principle deduced from the microscopic dynamics of the system. And this is indeed done successfully in some few simple cases. But in most cases, this precise dynamics is unknown, or the calculations are untractable. Therefore q can be obtained by fitting results with the analytic forms (q-exponentials, q-Gaussians) that emerge within nonextensive statistical mechanics. A third class of fitting parameters would be those which emerge from no specific theory, but are present in some convenient heuristic expressions. Within the q-statistical theory, we are not especially interested in those.

(vi) Dauxois considers "*particularly pressing*" an undoubtedly relevant question, namely "*does the q-Gaussian law describe the details of some physical problems?*". Many systems exist that have been shown to exhibit, at the available precision, q-Gaussians. A particularly instructive example concerns cold atoms in dissipative optical lattices. It was predicted in 2003 by Lutz (Lutz 2003) that these atoms should have a q-Gaussian distribution of velocities with

$$q = 1 + \frac{44E_R}{U_0}, \tag{1}$$

E_R and U_0 being respectively the microscopic *recoil energy* and *potential depth*. In 2006, an impressive confirmation was provided (Douglas et al. 2006), through both quantum Monte Carlo simulations, and real experiments with Cs atoms (the quality of the agreements are given by the correlation coefficients $R^2 = 0.995$ and $R^2 = 0.9985$ respectively). Other examples (some of them briefly mentioned in (Dauxois 2007)) that, within reasonably good approximation, have exhibited q-Gaussians are the motion of cells of *Hydra viridissima* (Upadhyaya et al. 2001), defect turbulence (Daniels et al. 2004), fluctuations of the magnetic field in the solar wind (Burlaga and Vinas 2004; Vinas et al. 2005; Burlaga and Vinas 2005; Burlaga et al. 2006; Burlaga et al. 2007; Burlaga et al. 2007), fluctuations of the temperature of the universe (Bernui et al. 2007), financial return distributions (Borland 2002a,b; Queiros 2005a,b; Queiros and Moyano 2006; Queiros et al. 2007; Queiros 2007), many-body Hamiltonian with long-range-interacting classical rotators approached through the HMF (Pluchino et al. 2007; Tsallis et al. 2007), logistic map at edge of chaos (Tirnakli et al. 2007, Tirnakli et al. 2009), granular matter (Arevalo et al. 2007), earthquakes (Caruso et al. 2007), etc.

(vii) "*More importantly*", as a second, "*particularly pressing*" question, Dauxois asks "*is anyone able to provide analytical predictions of the value of the q-index in terms of the microscopic parameters of the physical system?*". This is of course an important point, that any quantitatively mature statistical mechanical theory (and physical theory in general) is expected to satisfy. The Author might be unaware of the fact that many examples exist for which analytical calculations of q have been possible from microscopic or mesoscopic dynamics. Let us mention some of them. Of course, (1) already constitutes one such example. Another analytical result is the fact that the index q has been connected (Robledo 2005) with the standard critical exponent δ (the exponent characterizing the order parameter as a function of its thermodynamically conjugate variable, at a second-order critical phenomenon) as follows:

$$q = \frac{1+\delta}{2}. \tag{2}$$

In the area of quantum entanglement, the block entropy of fully entangled quantum magnetic chains involving only short-range interactions, has been recently proved to satisfy (Caruso and Tsallis 2007, Caruso and Tsallis 2008)

$$q = \frac{\sqrt{9+c^2}-3}{c}, \tag{3}$$

where c is the *central charge* emerging within conformal quantum field theory. This means that, for the Ising ferromagnet (i.e., $c = 1/2$) we have $q = \sqrt{37} - 6 = 0.08...$, and for the isotropic XY ferromagnet (i.e., $c = 1$), we have $q = \sqrt{10} - 3 = 0.16....$ In the limit $c \to \infty$ the BG result $q = 1$ is recovered. As another analytical connection, we may mention the Albert-Barabasi model in (Albert and Barabasi 2000) for (asymptotically) scale-free networks, for which we have (see (Thurner et al. 2007) for the connection of the Albert-Barabasi exponent and q)

$$q = \frac{2m(2-r)+1-p-r}{m(3-2r)+1-p-r}, \tag{4}$$

where (m, p, r) are microscopic parameters of the model. Let us also mention that, on the grounds of (Tsallis and Bukman 1996), the following connection is since long predicted between q and the anomalous diffusion space-time scaling exponent γ:

$$\gamma = \frac{2}{3-q}. \tag{5}$$

This prediction has been verified by now in various systems, such as the *Hydra viridissima* cells (Upadhyaya et al. 2001), defect turbulence (Daniels et al. 2004), α-XY inertial ferromagnetic model (Rapisarda and Pluchino 2005) (see (Anteneodo and Tsallis 1998) for details on the model), and granular matter flowing out from a silo (Arevalo et al. 2007).

As our last present illustration of analytical predictions involving q, we may mention one which is available since 1999 (Coraddu et al. 1999, 2002;

Lissia and Quarati 2005) and which has been successfully applied to plasma astrophysics (e.g., solar plasma). For weakly nonideal plasma, we have (Coraddu et al. 1999, 2002; Lissia and Quarati 2005)

$$|q-1| = 24\,\alpha^4\,\gamma^2 < 1\,, \tag{6}$$

where α (typically $0.4 < \alpha < 0.9$) is the *ion-ion correlation-function parameter* (see details in (Yan and Ichimaru 1986; Ichimaru 1992)), and $\gamma \equiv (Ze)^2 n^{1/3}/k_B T$ is the *plasma parameter* (see (7) of (Coraddu et al. 1999)), n being the *average density*.

(viii) The Author of (Dauxois 2007) makes, a few lines later, a short-cut which might mislead the quick reader. He states that "*Hilhorst and Schehr show that q-Gaussians do not pass a careful inspection.*". Instead, what these Authors have (interestingly) shown is that the *MTG* and *TMNT* models do not *pass* in the very specific sense that their limiting distributions are *not exactly* but only approximatively q-Gaussians. This point was already discussed in the beginning of the present manuscript.

(ix) The Author of (Dauxois 2007) states that "*if there is one lesson that has to be learned here, it is that one should be extremely careful when interpreting non-Gaussian data in terms of q-Gaussians*". This is certainly true. But, of course, this also is true for Gaussians, Lorentzians, circles, ellipses, straight lines (in fact, q-Gaussians are mere straight lines in $\ln_q y \equiv (y^{1-q}-1)/(1-q)$ versus x^2 representation), parabolas, etc, all the analytical tools with which physicists have attempted to interpret and quantitatively understand the plethora of experimental, observational and computational results they are dealing with since centuries and millenia. But of course, – *ce qui est clair sans le dire est encore plus clair en le disant!* –, *any* experimental, observational or computational setup can only provide a *finite* amount of numbers with *finite* precision. This will *always* allow for an *infinite* number of analytic curves or interpretations. The content of (Hilhorst and Schehr 2007) provides an interesting, and possibly fruitful, illustration of this obvious fact.

Acknowledgements Financial support by CNPq and Faperj (Brazilian Agencies) is gratefully acknowledged.

References

T. Dauxois, *Non-Gaussian distributions under scrutiny*, 0712.3153 [cond-mat.stat-mec]; J. Stat. Mech. N08001 (2007).

H.J. Hilhorst and G. Schehr, *A note on q-Gaussians and non-Gaussians in statistical mechanics*, J. Stat. Mech. P06003 (2007).

L.G. Moyano, C. Tsallis and M. Gell-Mann, *Numerical indications of a q-generalised central limit theorem*, Europhys. Lett. **73**, 813 (2006).

W. Thistleton, J.A. Marsh, K. Nelson and C. Tsallis (2006); now published in W. Thistleton, J.A. Marsh, K. Nelson and C. Tsallis, q-Gaussian approximants mimic non-extensive statistical-mechanical expectation for many-body probabilistic model with long-range correlations, Cent. Eur. J. Phys. **7**, 387 (2009).

C. Tsallis, *Possible generalization of Boltzmann-Gibbs statistics*, J. Stat. Phys. **52**, 479 (1988).

E.M.F. Curado and C. Tsallis, *Generalized statistical mechanics: connection with thermodynamics*, J. Phys. A **24**, L69 (1991); *Corrigenda:* **24**, 3187 (1991) and **25**, 1019 (1992).

C. Tsallis, R.S. Mendes and A.R. Plastino, *The role of constraints within generalized nonextensive statistics*, Physica A **261**, 534 (1998).

M. Gell-Mann and C. Tsallis, eds., *Nonextensive Entropy - Interdisciplinary Applications*, (Oxford University Press, New York, 2004).

J.P. Boon and C. Tsallis, eds., *Nonextensive Statistical Mechanics: New Trends, New Perspectives*, Europhys. News **36** (6) (European Physical Society, 2005); *Corrigendum:* Europhys. News **37** (1), 25 (2006).

S. Abe, H.J. Herrmann, P. Quarati, A. Rapisarda and C. Tsallis, eds. *Complexity, Metastability and Nonextensivity*, American Institute of Physics Conference Proceedings **965** (New York, 2007).

C. Tsallis, *Entropy, in Encyclopedia of Complexity and Systems Science* (11 volumes) (Springer, Berlin, 2009).

S. Umarov, C. Tsallis and S. Steinberg, On a q-central limit theorem consistent with nonextensive statistical mechanics, Milan J. Math. **76**, 307 (2008); C. Tsallis and S.M.D. Queiros, *Nonextensive statistical mechanics and central limit theorems I - Convolution of independent random variables and q-product*, in *Complexity, Metastability and Nonextensivity*, eds. S. Abe, H.J. Herrmann, P. Quarati, A. Rapisarda and C. Tsallis, American Institute of Physics Conference Proceedings **965**, 8 (New York, 2007); S.M.D. Queiros and C. Tsallis, *Nonextensive statistical mechanics and central limit theorems II - Convolution of q-independent random variables*, in *Complexity, Metastability and Nonextensivity*, eds. S. Abe, H.J. Herrmann, P. Quarati, A. Rapisarda and C. Tsallis, American Institute of Physics Conference Proceedings **965**, 21 (New York, 2007).

W. Thistleton, J.A. Marsh, K. Nelson and C. Tsallis, *Generalized Box-Muller method for generating q-Gaussian random deviates*, IEEE Trans. Inf. Theory **53**, 4805 (2007).

S. Umarov, C. Tsallis, M. Gell-Mann and S. Steinberg, Generalization of symmetric alpha-stable Levy distributions for q different from unity, submitted for publication (2009).

S. Umarov and C. Tsallis, *On multivariate generalizations of the q-central limit theorem consistent with nonextensive statistical mechanics*, in *Complexity, Metastability and Nonextensivity*, eds. S. Abe, H.J. Herrmann, P. Quarati, A. Rapisarda and C. Tsallis, American Institute of Physics Conference Proceedings **965**, 34 (New York, 2007); C. Vignat and A. Plastino, *Central limit theorem and deformed exponentials*, J. Phys. A **40**, F969 (2007).

A.R. Plastino and A. Plastino, *Non-extensive statistical mechanics and generalized Fokker-Planck equation*, Physica A **222**, 347 (1995).

C. Tsallis and D.J. Bukman, *Anomalous diffusion in the presence of external forces: exact time-dependent solutions and their thermostatistical basis*, Phys. Rev. E **54**, R2197 (1996).

C. Tsallis, *Nonextensive statistical mechanics, anomalous diffusion and central limit theorems*, Milan J. Math. **73**, 145 (2005).

R.J.V. dos Santos, *Generalization of Shannon's theorem for Tsallis entropy*, J. Math. Phys. **38**, 4104 (1997).

S. Abe, *Axioms and uniqueness theorem for Tsallis entropy*, Phys. Lett. A **271**, 74 (2000).

S. Abe, *Stability of Tsallis entropy and instabilities of Renyi and normalized Tsallis entropies: A basis for q-exponential distributions*, Phys. Rev. E **66**, 046134 (2002).

F. Topsoe, *Factorization and escorting in the game-theoretical approach to non-extensive entropy measures*, Physica A **365**, 91 (2006).

C. Anteneodo and C. Tsallis, *Multiplicative noise: A mechanism leading to nonextensive statistical mechanics*, J. Math. Phys. **44**, 5194 (2003).

F. Baldovin and A. Robledo, *Sensitivity to initial conditions at bifurcations in one-dimensional nonlinear maps: Rigorous nonextensive solutions*, Europhys. Lett. **60**, 518 (2002).

F. Baldovin and A. Robledo, *Universal renormalization-group dynamics at the onset of chaos in logistic maps and nonextensive statistical mechanics*, Phys. Rev. E **66**, R045104 (2002).
F. Baldovin and A. Robledo, *Nonextensive Pesin identity. Exact renormalization group analytical results for the dynamics at the edge of chaos of the logistic map.*, Phys. Rev. E **69**, 045202(R) (2004).
E. Mayoral and A. Robledo, *Tsallis' q index and Mori's q phase transitions at edge of chaos*, Phys. Rev. E **72**, 026209 (2005).
G. Wilk and Z. Wlodarczyk, *Interpretation of the nonextensivity parameter q in some applications of Tsallis statistics and Levy distributions*, Phys. Rev. Lett. **84**, 2770 (2000).
C. Beck, *Dynamical foundations of nonextensive statistical mechanics*, Phys. Rev. Lett. **87**, 180601 (2001).
C. Beck and E.G.D. Cohen, *Superstatistics*, Physica A **322**, 267 (2003).
E. Lutz, *Anomalous diffusion and Tsallis statistics in an optical lattice*, Phys. Rev. A **67**, 051402(R) (2003).
P. Douglas, S. Bergamini and F. Renzoni, *Tunable Tsallis distributions in dissipative optical lattices*, Phys. Rev. Lett. **96**, 110601 (2006).
A. Upadhyaya, J.-P. Rieu, J.A. Glazier and Y. Sawada, *Anomalous diffusion and non-Gaussian velocity distribution of Hydra cells in cellular aggregates*, Physica A **293**, 549 (2001).
K.E. Daniels, C. Beck and E. Bodenschatz, *Defect turbulence and generalized statistical mechanics*, Physica D **193**, 208 (2004).
L.F. Burlaga and A.F. Vinas, *Multiscale structure of the magnetic field and speed at 1 AU during the declining phase of solar cycle 23 described by a generalized Tsallis PDF*, J. Geophys. Res. Space Phys. **109**, A12107 (2004).
A.F. Vinas, R.L. Mace and R.F. Benson, *Dispersion characteristics for plasma resonances of Maxwellian and Kappa distribution plasmas and their comparisons to the IMAGE/RPI observations*, J. Geophys. Res. **110**, A06202 (2005).
L.F. Burlaga and A.F. Vinas, *Triangle for the entropic index q of non-extensive statistical mechanics observed by Voyager 1 in the distant heliosphere*, Physica A **356**, 375 (2005).
L.F. Burlaga, N.F. Ness and M.H. Acuna, *Multiscale structure of magnetic fields in the heliosheath*, J. Geophys. Res. Space Phys. **111**, A09112 (2006).
L.F. Burlaga, A. F-Vinas and C. Wang, *Tsallis distributions of magnetic field strength variations in the heliosphere: 5–90 AU*, J. Geophys. Res. Space Phys. **112**, A07206 (2007).
L.F. Burlaga, N.F. Ness, M.H. Acuna, *Magnetic fields in the heliosheath and distant heliosphere: Voyager 1 and 2 observations during 2005 and 2006*, Astrophys. J. **668**, 1246 (2007).
A. Bernui, C. Tsallis and T. Villela, *Deviation from Gaussianity in the cosmic microwave background temperature fluctuations*, Europhys. Lett. **78**, 19001 (2007).
L. Borland, *Closed form option pricing formulas based on a non-Gaussian stock price model with statistical feedback*, Phys. Rev. Lett. **89**, 098701 (2002).
L. Borland, *A theory of non-gaussian option pricing*, Quant. Finance **2**, 415 (2002).
S.M.D. Queiros, *On the emergence of a generalised Gamma distribution. Application to traded volume in financial markets*, Europhys. Lett. **71**, 339 (2005).
S.M.D. Queiros, *On non-Gaussianity and dependence in financial in time series: A nonextensive approach*, Quant. Finance **5**, 475–487 (2005).
J. de Souza, L.G. Moyano and S.M.D. Queiros, *On statistical properties of traded volume in financial markets*, Eur. Phys. J. B **50**, 165–168 (2006).
S.M.D. Queiros, L.G. Moyano, J. de Souza and C. Tsallis, *A nonextensive approach to the dynamics of financial observables*, Eur. Phys. J. B **55**, 161 (2007).
S.M.D. Queiros, *Are all highly liquid securities within the same class?*, Eur. Phys. J. B **60**, 265 (2007).
A. Pluchino, A. Rapisarda and C. Tsallis, *Nonergodicity and central limit behavior in long-range Hamiltonians*, Europhys. Lett. **80**, 26002 (2007. See also A. Pluchino, A. Rapisarda and C. Tsallis, A closer look at the indications of q-generalized Central Limit Theorem behavior in quasi-stationary states of the HMF model, Physica A 387, 3121 (2008), A. Figueiredo,

T.M. Rocha Filho and M.A. Amato, Ergodicity and central limit theorem in systems with long-range interactions, Eur. phys. Lett. **83**, 30011 (2008), A. Pluchino, A. Rapisarda and C. Tsallis, Comment on "Ergodicity and central limit theorem in systems with long-range interactions" by Figueiredo A. et al, Eur. phys. Lett. **85**, 60006 (2009), and A. Figueiredo, T.M. Rocha Filho and M.A. Amato, Reply to the Comment by A. Pluchino et al, Europhys. Lett. **85**, 60007 (2009).

C. Tsallis, A. Rapisarda, A. Pluchino and E.P. Borges, *On the non-Boltzmannian nature of quasi-stationary states in long-range interacting systems*, Physica A **381**, 143 (2007).

U. Tirnakli, C. Beck and C. Tsallis, *Central limit behavior of deterministic dynamical systems*, Phys. Rev. E **75**, 040106(R) (2007).

R. Arevalo, A. Garcimartin and D. Maza, *Anomalous diffusion in silo drainage*, Eur. Phys. J. E **23**, 191 (2007).

F. Caruso, A. Pluchino, V. Latora, S. Vinciguerra and A. Rapisarda, *Analysis of self-organized criticality in the Olami-Feder-Christensen model and in real earthquakes*, Phys. Rev. E **75**, 055101(R) (2007).

A. Robledo, *Unorthodox properties of critical clusters*, Mol. Phys. **103**, 3025 (2005).

F. Caruso and C. Tsallis, *Extensive nonadditive entropy in quantum spin chains*, in *Complexity, Metastability and Nonextensivity*, eds. S. Abe, H.J. Herrmann, P. Quarati, A. Rapisarda and C. Tsallis, American Institute of Physics Conference Proceedings **965**, 51 (New York, 2007).

R. Albert and A.-L. Barabasi, *Topology of evolving networks: Local events and universality*, Phys. Rev. Lett. **85**, 5234 (2000).

S. Thurner, F. Kyriakopoulos and C. Tsallis, *Unified model for network dynamics exhibiting nonextensive statistics*, Phys. Rev. E **76**, 036111 (2007).

A. Rapisarda and A. Pluchino, *Nonextensive thermodynamics and glassy behaviour in Hamiltonian systems*, Europhys. News **36**, 202 (2005) [Erratum: **37**, 25 (2006)].

C. Anteneodo and C. Tsallis, *Breakdown of the exponential sensitivity to the initial conditions: Role of the range of the interaction*, Phys. Rev. Lett. **80**, 5313 (1998).

M. Coraddu, G. Kaniadakis, A. Lavagno, M. Lissia, G. Mezzorani and P. Quarati, *Thermal distributions in stellar plasmas, nuclear reactions and solar neutrinos*, in *Nonextensive Statistical Mechanics and Thermodynamics*, eds. S.R.A. Salinas and C. Tsallis, Braz. J. Phys. **29**, 153 (1999).

M. Coraddu, M. Lissia, G. Mezzorani and P. Quarati, *Deuterium burning in Jupiter interior*, Physica A **305**, 282 (2002).

M. Lissia and P. Quarati, *Nuclear astrophysical plasmas: Ion distributions and fusion rates*, Europhys. News **36**, 211 (2005).

X.Z. Yan and S. Ichimaru, *Theory of interparticle correlations in dense, high-temperature plasmas. VI. Probability densities of the electric microfields*, Phys. Rev. A **34**, 2167 (1986).

S. Ichimaru, *Statistical Plasma Physics*, Vol. **1**, Paragr. 1.3 (Addison Wesley, Redwood City, 1992).

K.P. Nelson and S. Umarov, The relationship between Tsallis statistics, the Fourier transform, and nonlinear coupling (2008), 0811.3777 [cs.IT].

U. Tirnakli, C. Tsallis and C. Beck, *A closer look at time averages of the logistic map at the edge of chaos*, Phys. Rev. E **79**, 056209 (2009).

F. Caruso and C. Tsallis, *Nonadditive entropy reconciles the area law in quantum systems with classical thermodynamics*, Phys. Rev. E **78**, 021101 (2008).

Some Remarks on the Paper "On the q-type Distributions"

S.S. Nair and A. Kattuveettil

Abstract A claim is made in the paper Nadarajah and Kotz [On the q-type distributions, Physica A 377:465–468, (2007)] that the many q-densities, which are widely used in physics literature, are special cases of or associated with Burr distributions in classical statistical literature. In the present paper it is pointed out that the q-densities are not coming from Burr distributions or from other classical statistical distributions, and that q-distributions are extensions of the limiting forms for $q \to 1$. It is also shown that a statistical distribution which contains all q-distributions as special cases is the pathway model of Mathai [A pathway to matrix-variate gamma and normal densities, Linear Algebra and Its Applications 396:317–328, (2005)]. Tsallis statistics and superstatistics of Beck and Cohen [Superstatistics, Physica A 322:267–275, (2003)] are also examined here in the light of the discussions.

Keywords q-type distributions · Burr type distributions · Pathway model

1 Introduction

Nadarajah and Kotz (2007) claim that the q-type distributions, which are frequently used in physics literature, were available in statistical literature, for example, the Burr types III and XII distributions. The two Burr distributions quoted in the above paper have the following densities:

$$f_1(x) = kmx^{m-1}(1 + x^m)^{-k-1} \tag{1}$$

and

$$f_2(x) = kmx^{-m-1}(1 + x^{-m})^{-k-1}. \tag{2}$$

S.S. Nair (✉) and A. Kattuveettil
Centre for Mathematical Sciences, Pala Campus Arunapuram P.O, Pala-686 574, Kerala, India
e-mail: seemasuku@gmail.com; anithawilson82@yahoo.com

H.J. Haubold and A.M. Mathai (eds.), *Proceedings of the Third UN/ESA/NASA Workshop on the International Heliophysical Year 2007 and Basic Space Science*, Astrophysics and Space Science Proceedings, DOI 10.1007/978-3-642-03325-4_2,

In the present paper we show that a claim made in the paper Nadarajah and Kotz (2007) is misleading. It is pointed out that the q-densities are not coming from Burr distribution or from other classical statistical distributions.

2 Results

It can be seen that $f_1(x)$ in (1) is a power transformation on a type-2 beta density, namely

$$g_1(y) = c_1\, y^{\alpha-1}(1+y)^{-(\alpha+\beta)}, \quad \text{where } y = x^m \tag{3}$$

and c_1 is the normalizing constant, $\Re(\alpha) > 0$, $\Re(\beta) > 0$ where $\Re(\cdot)$ denotes the real part of $(\cdot)$. It is apparent that Burr had only considered a power transformation on y as in (3), which will be more versatile in model building and applications. Weibull did the same type of power transformation on exponential random variable, which produced a highly useful distribution known in the current statistical literature as Weibull distribution. The density in (2) is obtained from the fact that x and $\frac{1}{x}$ belong to the same family of distributions when x is distributed as a type-2 beta or generalized gamma or stretched exponential distributions. It is quite obvious that (1) and (2) have no connections to q-type distributions. If Burr had taken the density in the form

$$f_3(x) = c_3\, x^{\gamma-1}(1+ax)^{-\beta} \tag{4}$$

with $a > 0,\ \gamma > 0,\ \beta - \gamma > 0$ and c_3 being the normalizing constant then one could have stretched the imagination and say that if a is replaced by $b(q-1)$ with $b > 0,\ q > 1$ and β by $\frac{1}{q-1}$ then one could obtain a q-type distribution from (4). As long as there is no coefficient $a > 0$ present in the model there is no connection to q-type distribution from (1).

Equations (3) and (4) of Nadarajah and Kotz (2007) are type-1 beta forms which cannot come from Burr distributions in (1) or (2), which are type-2 beta forms, unless one goes through a transformation. If Burr had taken a form

$$f_4(x) = c_4\, x^{\gamma-1}(1-ax)^b, \quad 1-ax > 0, a > 0, b > 0 \tag{5}$$

with c_4 being normalizing constant then by taking $a = a_1(1-q)$ and $b = \frac{1}{1-q}$ with $q < 1,\ a_1 > 0$ one could go into a q-type distribution from (5). This is not possible as long as $a > 0$ is not there as a coefficient of x in $1 - ax > 0$. Thus forms such as (4) and (5) above can be converted to q-type distributions and Burr distributions have nothing to do with the forms in (4) and (5) above.

Type-1 beta forms are given (3), (4), and (6) of Nadarajah and Kotz (2007) and the statements therein are quite misleading. It may be pointed out that a type-1 beta form is not available by the compounding procedure in (5) of Nadarajah and Kotz (2007), by taking a gamma type conditional density and a gamma type prior density for the shape parameter. Let the conditional density of x, given the shape parameter β, be given by

$$f_5(x|\beta) = \frac{\delta\beta^{\frac{\alpha}{\delta}}}{\Gamma\left(\frac{\alpha}{\delta}\right)}\, x^{\alpha-1}\mathrm{e}^{-\beta x^{\delta}},\ \ \beta > 0,\ \frac{\alpha}{\delta} > 0,\ x > 0, \tag{6}$$

and let the prior density of β be given by

$$g(\beta) = \frac{a^{\gamma}}{\Gamma(\gamma)}\, \beta^{\gamma-1}\mathrm{e}^{-a\beta},\ \ a > 0,\ \gamma > 0,\ \beta > 0. \tag{7}$$

Then the unconditional density of x is given by

$$\begin{aligned} f_6(x) &= \int_0^{\infty} f_5(x|\beta)g(\beta)\mathrm{d}\beta \\ &= \frac{\delta a^{\gamma}}{\Gamma\left(\frac{\alpha}{\delta}\right)\Gamma(\gamma)}\, x^{\alpha-1} \int_0^{\infty} \beta^{\frac{\alpha}{\delta}+\gamma-1}\mathrm{e}^{-\beta(a+x^{\delta})}\mathrm{d}\beta \ \ \text{for } a > 0,\ \frac{\alpha}{\delta} > 0, \\ &\qquad \gamma > 0,\ \delta > 0. \end{aligned}$$

But for the integral to be convergent $(a + x^{\delta})$ must be > 0 and the integral gives

$$f_6(x) = \frac{\delta a^{-\frac{\alpha}{\delta}}\Gamma\left(\frac{\alpha}{\delta}+\gamma\right)}{\Gamma\left(\frac{\alpha}{\delta}\right)\Gamma(\gamma)}\, x^{\alpha-1}\left(1+\frac{x^{\delta}}{a}\right)^{-\left(\frac{\alpha}{\delta}+\gamma\right)},\ \text{for } x > 0,\ a > 0,\ \gamma > 0,\ \frac{\alpha}{\delta} > 0. \tag{8}$$

Observe that only a form of the type $(1 + \frac{x^{\delta}}{a})^{-\eta}$ where $1 + \frac{x^{\delta}}{a} > 0,\ a > 0,\ x^{\delta} > 0$, $\eta > 0$, that is, only a type-2 beta form can come from such a consideration. In other words a type-1 beta form cannot come because for the convergence of the integral $a + x^{\delta}$ must be positive with $a > 0$ and $x^{\delta} > 0$. The discussions in Nadarajah and Kotz (2007) gives the impression that a type-1 beta form is available from the unconditional density coming from (6) to (8) above. It is to be pointed out here that the superstatistics of Beck and Cohen (2003) and Beck (2006) are available from the procedure in (6) to (8) above. It goes without saying that only type-2 beta form or a q-density of the form

$$f_7(x) = c_7\, x^{\alpha-1}[1 + a(q-1)x^{\delta}]^{-\frac{1}{q-1}}\ \ \text{for } a > 0,\ q > 1 \tag{9}$$

is available from superstatistics considerations. It may be noted that much more general forms for $f_5(x|\beta)$ and $g(\beta)$ can be considered. This is illustrated in Mathai and Haubold (2007). Also one could look for more general compatible densities for $f_5(x|\beta)$ and $g(\beta)$ and then look at the unconditional or posterior density to see whether type-1 beta form is available from such a process.

Burr density as given in (1) or (2) as well as any other density in statistical literature gives only one functional form. For various values of the parameters therein one has a whole family of densities but all members of this family are within the same functional form. The Burr density in (1) or in (2) has two parameters k and m. By taking different values for k and m one gets different members within the family. But the functional form remains the same. Whatever be the values of k and m,

(1) remains as a generalized type-2 beta form. No parameter value, $m = 1$ or any other value, will make the type-2 beta form change into an exponential form or any other form. But q-densities of Tsallis or Beck and Cohen are of the type that at the parameter value $q = 1$ the functional form thereof changes from type-2 beta form to exponential form so that physical interpretation can be given to this phenomenon. If $q = 1$ is the stable situation then the type-2 beta form for $q \neq 1,\ q > 1$ can describe unstable or chaotic situations. This is the distinction between a q-density in recent physics literature and classical statistical densities, however generalized they may be. The only statistical density which switches from one functional form to another, in fact three different functional forms, through a pathway parameter therein, is the pathway model of Mathai (2005). If the exponential form is the stable solution then the chaotic forms are given by $q \neq 1$ or the stable form is extended to include many unstable situations for $q \neq 1$. Through q-densities one can go from extensivity to nonextensivity in statistical mechanics. The q-densities in physics literature originated from the paper of Tsallis (1988). The many q-distributions referred in the reference list of Nadarajah and Kotz (2007) are beautiful results on their own rights but are not connected to Burr distributions or other classical statistical distributions.

It appears that the only statistical distributions connecting to all forms of q-densities in physics literature are the pathway models of Mathai (2005), where the scalar version of the pathway density is given as follows:

$$f_8(x) = c_8\ |x|^{\gamma}\ [1 - a(1-\alpha)|x|^{\delta}]^{\frac{\eta}{1-\alpha}}$$
$$\text{for}\quad a > 0,\ \delta > 0,\ \eta > 0, 1 - a(1-\alpha)x^{\delta} > 0,\quad -\infty < x < \infty. \tag{10}$$

Observe that for $\alpha < 1$ it is a finite range density with $1 - a(1-\alpha)|x|^{\delta} > 0$ and for $\alpha < 1$, (10) will remain in the generalized type-1 beta family. For $\alpha > 1$, writing $1 - \alpha = -(\alpha - 1)$ we have the form

$$f_9(x) = c_9\ |x|^{\gamma}\ [1 + a(\alpha-1)|x|^{\delta}]^{-\frac{\eta}{\alpha-1}},\quad a > 0,\ \alpha > 1,\ \eta > 0 \tag{11}$$

and this form is the generalized type-2 beta form. For $\alpha \to 1$ both (10) and (11) will go to the exponential form, from the fact that

$$\begin{aligned}\lim_{\alpha\to 1}[1 - a(1-\alpha)|x|^{\delta}]^{\frac{\eta}{1-\alpha}} &= \lim_{\alpha\to 1}[1 + a(\alpha-1)|x|^{\delta}]^{-\frac{\eta}{\alpha-1}}\\ &= \mathrm{e}^{-a\eta|x|^{\delta}}.\end{aligned} \tag{12}$$

Thus α here is called the pathway parameter taking the family of densities to three different functional forms in (10), (11), and (12). It is shown in Mathai (2005) that almost all continuous distributions in current use are either available as particular cases of (10) or from simple transformations in (10). If a connection is to be established between the many q-densities in physics literature to some statistical distributions then such a statistical distribution is the pathway model of Mathai (2005) as given in (10). Observe that (10) contains Tsallis statistics as well as superstatistics of Beck and Cohen. It may be pointed out also that for $\gamma = 0$ in (10), Tsallis statistics is more general than the superstatistics because Tsallis statistics can cover all the forms $\alpha < 1, \alpha > 1$ and $\alpha \to 1$ with $\gamma = 0$ in (10) whereas superstatistics can cover only the cases $\alpha > 1$ and $\alpha \to 1$. But superstatistics considerations can produce the factor of the type $|x|^{\gamma}$ with $\gamma \neq 0$ whereas Tsallis considerations can produce only models with $\gamma = 0$ but all cases $\alpha < 1, \alpha > 1$ and $\alpha \to 1$. It is to be pointed out here that in (10) when $\alpha > 1$ one can replace $|x|^{\delta}$ by $|x|^{-\delta}$ because in this case both x and $\frac{1}{x}$ will belong to the same family of distributions.

It is also pointed out in Mathai and Haubold (2007) that a pathway model of the form in (10) is also available by optimizing a generalized entropy. This generalized entropy, encompassing Tsallis entropy and many other entropy measures of the same class are variants of the classical generalized entropy of order α of Havrda-Charvát, the details are also given in Mathai and Haubold (2007).

3 Conclusion

In conclusion, one may say, as pointed out earlier, that the q-densities in physics literature are not connected to Burr distributions or such other distributions in classical statistical literature and the q-distributions are not special cases of such statistical distributions. The q-densities in physics literature are beautiful extensions of the classical results in the respective topics and they are not particular cases of Burr distribution or other classical statistical distributions. The only statistical density which contains all types of q-densities, Tsallis statistics, superstatistics etc is the pathway density of Mathai (2005).

Acknowledgment The authors would like to thank the Department of Science and Technology, Government of India, New Delhi, for the financial assistance for this work under project No. SR/S4/MS:287/05 and the Centre for Mathematical Sciences for providing all facilities.

References

Nadarajah, S., Kotz, S.: On the q-type distributions. Physica A **377**, 465–468 (2007).
Beck, C., Cohen, E.G.D.: Superstatistics. Physica A **322**, 267–275 (2003).
Beck, C.: Stretched exponentials from superstatistics. Physica A **365**, 96–101 (2006).
Mathai, A.M., Haubold, H.J.: Pathway model, superstatistics, Tsallis statistics, and a generalized measure of entropy. Physica A **375**, 110–122 (2007).
Tsallis, C.: Possible generalizations of Boltzmann-Gibbs statistics. Journal of Statistical Physics **52**, 479–487 (1988).
Mathai, A.M.: A pathway to matrix-variate gamma and normal densities. Linear Algebra and Its Applications **396**, 317–328 (2005).

Canonical Formulation of Fractional Kinetics

Sumiyoshi Abe

Abstract A canonical formalism is developed for a fractional diffusion equation, introducing an auxiliary field. The equation is recast in the form of the canonical equation of motion with respect to the Hamiltonian and the Dirac bracket. The generator of a dilatation transformation is constructed, and then the scale invariance and its violation are discussed.

Keywords Fractional diffusion equation · Scale invariance · Canonical formalism

Fractional kinetics has been attracting continuous attention for more than a couple of decades (Le Mehaute 1984; Hilfer 2000; Metzler and Klafter 2000; Zaslavsky 2005). This theory offers a consistent mathematical framework for describing exotic phenomena such as anomalous diffusion observed in complex systems.

Consider the following fractional diffusion equation in a single spatial dimension:

$$\frac{\partial f(x,t)}{\partial t} = D\frac{\partial^{\gamma} f(x,t)}{\partial |x|^{\gamma}}, \quad (-\infty < x < \infty,\ 0 \le t,\ 0 < \gamma < 2), \tag{1}$$

where $f(x,t)dx$ is the probability of finding a diffusing particle in the interval, $x \sim x+dx$, at time t and D is a generalized diffusion constant. The Riesz representation of the fractional Laplacian, which appears on the right-hand side, is defined by (Zaslavsky 2005)

$$\frac{d^{\gamma} g(x)}{d|x|^{\gamma}} = -\frac{1}{K(\gamma)}\int_{0+}^{\infty}\frac{dy}{y^{\gamma+1}}[g(x-y) - 2g(x) + g(x+y)], \tag{2}$$

S. Abe (✉)
Department of Physical Engineering, Mie University, Mie 514–8507, Japan
and
Institut Supérieur des Matériaux et Mécaniques Avancés, 44 F. A. Bartholdi, 72000 Le Mans, France

H.J. Haubold and A.M. Mathai (eds.), *Proceedings of the Third UN/ESA/NASA Workshop on the International Heliophysical Year 2007 and Basic Space Science*, Astrophysics and Space Science Proceedings, DOI 10.1007/978-3-642-03325-4_3,

with $K(\gamma) = 2\Gamma(-\gamma)\cos(\pi\gamma/2)$ for $\gamma \neq 1$ and $K(1) = -\pi$, where $\Gamma(s)$ is Euler's gamma function. This is a regularized form of the operator

$$\frac{d^\gamma}{d|x|^\gamma} = -\frac{1}{2\cos(\pi\gamma/2)}\left[\frac{d^\gamma}{dx^\gamma} + \frac{d^\gamma}{d(-x)^\gamma}\right]. \tag{3}$$

Using the Fourier transformation, $\tilde{g}(k) = \int_{-\infty}^{\infty} dx\, e^{ikx} g(x)$, one has the following correspondence relation:

$$\frac{d^\gamma g(x)}{d|x|^\gamma} \leftrightarrow -|k|^\gamma \tilde{g}(k). \tag{4}$$

The solution of (1) with the initial condition, $f(x,0) = \delta(x)$, is found to be (Abe and Thurner 2005; and the references therein)

$$f(x,t) = t^{-1/\gamma} L_\gamma\left(x/t^{1/\gamma}\right). \tag{5}$$

Here, $L_\gamma(x)$ is the symmetric Lévy stable distribution (Gnedenko and Kolmogorov 1968; Feller 1971), which is given by

$$L_\gamma(x) = \frac{1}{2\pi}\int_{-\infty}^{\infty} dk \exp\left(-ikx - D|k|^\gamma\right). \tag{6}$$

Except the Gaussian limit ($\gamma \to 2-0$), this distribution decays as a power law: $L_\gamma(x) \sim |x|^{-1-\gamma}$ for a large value of $|x|$, implying asymptotic scale invariance of the system. It is the limit distribution of independent and identically-distributed (i.i.d.) random variables with divergent second moments characterized by the Lévy-Gnedenko generalized central limit theorem (Gnedenko and Kolmogorov 1968; Feller 1971). From the scaling property in (5), it can be seen that spreading of the distribution (defined, e.g., by the half width) grows as $\sim t^{1/\gamma}$. Since $0 < \gamma < 2$, the system diffuses faster than normal diffusion $\left(\sim t^{1/2}\right)$, and this is why the phenomenon is termed superdiffusion.

As mentioned above, the Lévy distribution is scale invariant only asymptotically. Such a symmetry is violated in the regime of small $|x|$.

In this article, we develop a canonical formalism for the fractional diffusion equation in (1), introducing an auxiliary field and employing Dirac's method of singular Lagrangians. Constructing the generator of dilatation transformation, we also discuss how the exact dilatation symmetry is violated for the system Hamiltonian. The present work can be viewed as a fractional generalization of a recent one on the canonical formulation of the Fokker-Planck equation (Abe 2004). It may present a step toward establishing Hamiltonian variational principle for fractional kinetics.

The fractional diffusion equation in (1) is of the first order in time. Accordingly, to develop a canonical formalism, we introduce an auxiliary field, denoted by

$\Lambda(x,t)$, which is not necessarily a probability distribution itself. With it, we construct the following action functional:

$$I = \int_{0}^{\infty} dt \int_{-\infty}^{\infty} dx\, \pounds, \tag{7}$$

where $\pounds$ is the Lagrangian density given by

$$\pounds = \frac{1}{2}\left(\Lambda \frac{\partial f}{\partial t} - \frac{\partial \Lambda}{\partial t} f\right) - D\Lambda \frac{\partial^{\gamma} f}{\partial |x|^{\gamma}}. \tag{8}$$

Using the formula for fractional integration by part

$$\int_{-\infty}^{\infty} dx\, g(x) \frac{d^{\gamma} h(x)}{d|x|^{\gamma}} = \int_{-\infty}^{\infty} dx \frac{d^{\gamma} g(x)}{d|x|^{\gamma}} h(x) \tag{9}$$

with the assumption that the surface term vanishes, we can rewrite the action as follows:

$$I = -\int_{0}^{\infty} dt \left\langle \frac{\partial \Lambda}{\partial t} + D \frac{\partial^{\gamma} \Lambda\,(x,t)}{\partial |x|^{\gamma}} \right\rangle, \tag{10}$$

where $\langle Q \rangle \equiv \int_{-\infty}^{\infty} dx\, Q(x,t) f(x,t)$.

The variations of the action in (7) with respect to f and Λ lead to

$$\frac{\partial \Lambda(x,t)}{\partial t} = -D \frac{\partial^{\gamma} \Lambda(x,t)}{\partial |x|^{\gamma}}, \tag{11}$$

$$\frac{\partial f(x,t)}{\partial t} = D \frac{\partial^{\gamma} f(x,t)}{\partial |x|^{\gamma}}, \tag{12}$$

respectively. Equation (12) is precisely equal to (1). It is of interest to observe that these two equations are formally related to each other through time reversal.

The canonical momenta conjugate to f and Λ are calculated to be

$$\Pi_f = \frac{\partial \pounds}{\partial(\partial f/\partial t)} = \frac{\Lambda}{2}, \tag{13}$$

$$\Pi_\Lambda = \frac{\partial \pounds}{\partial(\partial \Lambda/\partial t)} = -\frac{f}{2}. \tag{14}$$

Since $\partial f/\partial t$ and $\partial \Lambda/\partial t$ cannot be solved in terms of Π_f and Π_Λ, (13) and (14) give rise to the following set of primary constraints (Dirac 2001):

$$\chi_1 = \Pi_f - \frac{\Lambda}{2} \approx 0, \tag{15}$$

$$\chi_2 = \Pi_\Lambda + \frac{f}{2} \approx 0, \tag{16}$$

where the symbol "$\approx$" denotes the weak equality in Dirac's notation.

The basic equal-time Poisson bracket relations read

$$\{f(x,t), \Pi_f(x',t)\} = \delta(x-x'), \tag{17}$$

$$\{\Lambda(x,t), \Pi_\Lambda(x',t)\} = \delta(x-x'). \tag{18}$$

To examine the consistency of the constraints in (15) and (16), their Poisson bracket should be examined. Using (17) and (18), we obtain

$$\{\chi_1(x,t), \chi_2(x',t)\} = -\delta(x-x'). \tag{19}$$

This quantity does not identically vanish. Accordingly, (15) and (16) are the second-class constraints in Dirac's terminology (Dirac 2001). Actually, appearance of the second-class constraints is common in systems of the first order in time.

To eliminate these second-class constraints, let us introduce the Dirac bracket. For general physical quantities, $A(t)$ and $B(t)$, which are functionals of f and Λ, the Dirac bracket is defined by

$$\{A(t), B(t)\}^* = \{A(t), B(t)\}$$
$$- \sum_{i,j=1}^{2} \int_{-\infty}^{\infty} \int_{-\infty}^{\infty} dx\, dx' \{A(t), \chi_i(x,t)\} C_{ij}(x,x') \{\chi_j(x',t), A(t)\}. \tag{20}$$

Here, $C_{ij}(x,x')$ is a quantity satisfying

$$\sum_{k=1}^{2} \int_{-\infty}^{\infty} dx'' \{\chi_i(x,t), \chi_k(x'',t)\} C_{kj}(x'',x') = \delta_{ij}\delta(x-x'). \tag{21}$$

An explicit calculation shows that

$$C_{11}(x,x') = C_{22}(x,x') = 0, \tag{22}$$

$$C_{12}(x,x') = -C_{21}(x,x') = \delta(x-x'). \tag{23}$$

With the Dirac bracket, the commutation relations between those second-class constraints and an arbitrary physical quantity identically vanish, and thus the constraints are now regarded as the identities.

The Hamiltonian is constructed as usual by the Legendre transformation:

$$
\begin{aligned}
H &= \int_{-\infty}^{\infty} dx \left(\Pi_f \frac{\partial f}{\partial t} + \Pi_\Lambda \frac{\partial \Lambda}{\partial t} - \pounds \right) \\
&= \int_{-\infty}^{\infty} dx D \frac{\partial^\gamma \Lambda}{\partial |x|^\gamma} f \\
&= \int_{-\infty}^{\infty} dx D \frac{\partial^{\gamma/2} \Lambda}{\partial |x|^{\gamma/2}} \frac{\partial^{\gamma/2} f}{\partial |x|^{\gamma/2}}.
\end{aligned}
\tag{24}
$$

Using this Hamiltonian, now we can derive the following canonical equations of motion:

$$
\begin{aligned}
\frac{\partial \Lambda (x,t)}{\partial t} &= \{\Lambda (x,t), H\}^* \\
&= -D \frac{\partial^\gamma \Lambda (x,t)}{\partial |x|^\gamma},
\end{aligned}
\tag{25}
$$

$$
\begin{aligned}
\frac{\partial f(x,t)}{\partial t} &= \{f(x,t), H\}^* \\
&= D \frac{\partial^\gamma f(x,t)}{\partial |x|^\gamma},
\end{aligned}
\tag{26}
$$

which correctly reproduce (11) and (12), respectively. This establishes the canonical formalism for the fractional diffusion equation.

As discussed earlier, the fractional diffusion equation has as a solution the Lévy distribution in (5), which decays as a power law, being indivisibly related to the superdiffusion phenomenon. However, the power law appears only asymptotically, and the exact scale invariance in the whole range of x is violated. In what follows, we discuss this point based on the canonical formalism developed above.

The generator of the dilatation transformation reads (Abe 2004)

$$
G = -2 \int_{-\infty}^{\infty} dx\, x \frac{\partial \Lambda}{\partial x} \Pi_\Lambda = \int_{-\infty}^{\infty} dx\, x \frac{\partial \Lambda}{\partial x} f. \tag{27}
$$

This quantity turns out to yield the following Dirac-bracket relations:

$$
\{G(t), \Lambda (x,t)\}^* = x \frac{\partial \Lambda (x,t)}{\partial x}, \tag{28}
$$

$$
\{G(t), f(x,t)\}^* = \frac{\partial}{\partial x} [x f(x,t)]. \tag{29}
$$

Accordingly, the finite transformations are given by

$$e^{\{(\ln\lambda)G(t)\}^*}\Lambda(x,t)e^{-\{(\ln\lambda)G(t)\}^*} = e^{(\ln\lambda)x\partial/\partial x}\Lambda(x,t) = \Lambda(\lambda x,t), \tag{30}$$

$$e^{\{(\ln\lambda)G(t)\}^*}f(x,t)e^{-\{(\ln\lambda)G(t)\}^*} = e^{(\ln\lambda)(\partial/\partial x)x}f(x,t) = \lambda f(\lambda x,t), \tag{31}$$

where λ is a positive constant scale factor, and the notation, $e^{\{A\}^*}B\,e^{-\{A\}^*} = B + \{A,B\}^* + (1/2!)\left\{A,\{A,B\}^*\right\}^* + \cdots$, is used. Note that the normalization condition on f is preserved, $\int_{-\infty}^{\infty}dx\,\lambda f(\lambda x,t) = \int_{-\infty}^{\infty}dx\,f(x,t) = 1$, whereas the auxiliary field does not have to be normalized, since it is not a probability distribution.

Now, the system is scale invariant if G is the generator of a good symmetry:

$$\{G,H\}^* = 0. \tag{32}$$

A straightforward calculation shows

$$\{G,H\}^* = D\int_{-\infty}^{\infty}dx\,x\left[\frac{\partial\Lambda}{\partial x}\frac{\partial^\gamma f}{\partial|x|^\gamma} - f\frac{\partial^\gamma}{\partial|x|^\gamma}\left(\frac{\partial\Lambda}{\partial x}\right)\right], \tag{33}$$

provided that $[\partial/\partial x, \partial^\gamma/\partial|x|^\gamma] = 0$ has been used. Performing integration by part and dropping the surface terms, we have

$$\{G,H\}^* = D\int_{-\infty}^{\infty}dx\frac{\partial\Lambda}{\partial x}\left[x\frac{\partial^\gamma f}{\partial|x|^\gamma} - \frac{\partial^\gamma(xf)}{\partial|x|^\gamma}\right], \tag{34}$$

where we have employed the formula in (9). Using (2), the second term in the integrand is rewritten as

$$\frac{\partial^\gamma[xf(x,t)]}{\partial|x|^\gamma} = x\frac{\partial^\gamma f(x,t)}{\partial|x|^\gamma} - \frac{1}{K(\gamma)}\int_{0+}^{\infty}\frac{dy}{y^\gamma}[f(x+y,t) - f(x-y,t)]. \tag{35}$$

Therefore, Equation (34) also has the following form:

$$\{G,H\}^* = -\frac{D}{K(\gamma)}\int_{-\infty}^{\infty}dx\,\Lambda(x,t)\int_{0+}^{\infty}\frac{dy}{y^\gamma}\frac{\partial}{\partial x}[f(x+y,t) - f(x-y,t)]. \tag{36}$$

Equation (34) [or, equivalently (36)] explicitly shows how the exact dilatation symmetry in the whole range of x expressed in (32) is violated in the system.

To summarize, we have developed a canonical formalism for the fractional diffusion equation and recast the equation in the form of the canonical equation of motion with respect to the Hamiltonian and the Dirac bracket. The formalism has been used to describe how the exact dilatation symmetry is violated. The variational principle for ordinary kinetic theory has been developed in the literature (Éboli 1988) based on the Gaussian Ansatz for distributions. There, the existence of a finite second moment of the basic random variable plays an essential role. The present discussion can be seen as a step toward the formulation of a variational principle for fractional kinetics, where distributions tend to have divergent second moments. A trial function may have a fat tail, in general.

Acknowledgement This work was supported in part by the Grant-in-Aid for Scientific Research (B) of the Ministry of Education.

References

Abe S.: Phys. Rev. E **69**, 016102 (2004)
Abe S., Thurner, S.: Physica A **356**, 403 (2005)
Dirac P. A. M.: Lectures on Quantum Mechanics. Dover, New York (2001)
Éboli O., Jackiw R., Pi S.-Y.: Phys. Rev. D **37**, 3557 (1988)
Feller W.: An Introduction to Probability Theory and Its Applications, Vol. II. Wiley, New York (1971)
Gnedenko B. V., Kolmogorov A. N.: Limit Distributions for Sums of Independent Random Variables. Addison-Wesley, Reading, Massachusetts (1968)
Hilfer R. (ed): Applications of Fractional Calculus in Physics. World Scientific, Singapore (2000)
Le Mehaute A.: J. Stat. Phys. **36**, 665 (1984)
Metzler R., Klafter J.: Phys. Rep. **339**, 1 (2000)
Zaslavsky G. M.: Hamiltonian Chaos and Fractional Dynamics. Oxford University Press, Oxford (2005)

Out-of-Equilibrium Statistical Mechanics in a Hamiltonian System with Mean-Field Interaction

Yoshiyuki Y. Yamaguchi

Abstract Systems with finite degrees of freedom and with long-range interaction are frequently trapped at quasi-stationary states before relaxing to thermal equilibrium. Short-time relaxation to quasi-stationary states is approximated by the Vlasov equation, and a statistical theory based on the Vlasov description is introduced and applied to the Hamiltonian mean-field model. The theory predicts a one-body distribution for a quasi-stationary state from a given waterbag initial distribution, and a critical curve is described on a two-dimensional parameter plane which represents a family of waterbag initial distributions. The critical curve divides the parameter plane into magnetized and non-magnetized phases of quasi-stationary states. The theoretical prediction is checked by comparing with a numerically obtained critical curve for systems with finite degrees of freedom.

Keywords Quasi-stationary state · Out-of-equilibrium statistics · Long-range interaction

1 Introduction

In nature, long-range interactions (Dauxois et al. 2002) appear in many Hamiltonian systems like a self-gravitating system, a plasma system, a vortex system in a Euler fluid and so on. In a such system the state is frequently trapped in the so-called quasi-stationary state (QSS) before relaxing to thermal equilibrium, and the trapping time diverges as number of particles increases. Due to the long lifetime of QSSs, the dynamically accessible regimes are solely QSSs in a system consisting of a large number of particles subject to long-range couplings. This fact motivates us to construct statistical mechanics for QSSs.

Y.Y. Yamaguchi (✉)
Department of Applied Mathematics and Physics, Kyoto University, 606-8501, Kyoto, Japan
e-mail: yyama@amp.i.kyoto-u.ac.jp

H.J. Haubold and A.M. Mathai (eds.), *Proceedings of the Third UN/ESA/NASA Workshop on the International Heliophysical Year 2007 and Basic Space Science*, Astrophysics and Space Science Proceedings, DOI 10.1007/978-3-642-03325-4_4,

Dynamics of a Hamiltonian system composed by N particles is governed by the Liouville equation for N-body distribution, and the Liouville equation is reduced to the Boltzmann equation for one-body distribution through the BBGKY hierarchy (Balescu 1997). The Boltzmann equation has a collisional term which includes two-body distribution, and hence it is not closed. If the collisional term can be neglected, one obtains the Vlasov equation, which is a closed equation of the one-body distribution.

In context of the one-body distribution, a scenario of relaxation in Hamiltonian systems with long-range interactions is described as follows (Lynden-Bell 1967; Yamaguchi et al. 2004; Barré et al. 2006). An initial state relaxes to QSS through violent relaxation where the Vlasov approximation is good enough. After reaching QSS, the state is in the collisional relaxation where the Boltzmann equation must be applied. QSS is realized as the final state of the violent relaxation, and hence statistical mechanics based on the Vlasov equation must be useful to describe QSS.

Inspired by the pioneering work of Lynden-Bell (1967) for self-gravitating systems, a statistical theory for QSS is applied to the Hamiltonian mean-field (HMF) model (Antoniazzi et al. 2007a). The theory predicts one-body distributions in QSSs from initial distributions without any free parameters, and the predicted distributions are in good agreement with numerically obtained ones. Moreover, on a two-dimensional parameter plane, a tricritical point is discovered where first and second order phase transition lines collapse (Antoniazzi et al. 2007b). It is worth remarking that QSSs show richer critical phenomena than thermal equilibrium, where the HMF model has a second order phase transition only (Antoni and Ruffo, 1995).

The purpose of this article is to introduce the statistical theory for QSSs, and to numerically confirm its validity. The validity is partly reported in the previously paper (Antoniazzi et al. 2007a, b), but in this article, we explore the whole of the two-dimensional parameter space mentioned above.

2 Model

We introduce the HMF model (Inagaki 1993; Antoni and Ruffo 1995), which is simple but shares similarities with self-gravitating sheet model (Tsuchiya et al. 1994) and charged sheet model (Elskens and Escande 2003), which have long-range interactions. The HMF model is a model of ferromagnetic body, and is expressed by the following Hamiltonian

$$H = \frac{1}{2}\sum_{j=1}^{N} p_j^2 + \frac{1}{2N}\sum_{j,k=1}^{N} [1 - \cos(\theta_j - \theta_k)]. \tag{1}$$

The system consists of N rotators confined to the unit circle, and θ_j and p_j represent angle and its conjugate momentum of the j-th rotator respectively. The factor $1/N$ is added in the second term of the right-hand-side to recover extensivity, but it can be removed by rescaling the time scale.

An order parameter of this system is magnetization, and modulus M and phase α of magnetization are defined by the following equation

$$Me^{i\alpha} = \frac{1}{N}\sum_{j=1}^{N} e^{i\theta_j}.$$

The Hamiltonian is rewritten by using magnetization as

$$H = \sum_{j=1}^{N}\left[\frac{p_j^2}{2} + \frac{1 - M\cos(\theta_j - \alpha)}{2}\right],$$

and hence energy of a rotator at (θ, p) is read as

$$\frac{p^2}{2} + \frac{1 - M(t)\cos\theta}{2}. \tag{2}$$

We assumed $\alpha = 0$ without loss of generality from the rotational symmetry of the system, and we call M magnetization for simplicity.

The statistical theory introduced in Sec. 3 is applied for waterbag initial distributions, which are expressed by two-value functions on μ-space, (θ, p)-plane. The waterbag initial distribution is expressed as

$$f(\theta, p, t = 0) = \begin{cases} f_0 & (\theta, p) \in D \\ 0 & \text{otherwise} \end{cases}, \tag{3}$$

where $f(\theta, p, t)$ is the one-body probability distribution function at time t. We choose the domain D as a rectangle

$$D = \{(\theta, p) \in \mu \mid |\theta| \le \Delta\theta,\ |p| \le \Delta p\}, \tag{4}$$

where $0 \le \Delta\theta \le \pi$ and $\Delta p \ge 0$. Satisfying the probability constraint

$$\int_{-\pi}^{\pi} d\theta \int_{-\infty}^{\infty} dp\ f(\theta, p, t = 0) = 1,$$

the factor f_0 is determined as $f_0 = 1/4\Delta\theta\Delta p$.

The pair of $\Delta\theta$ and Δp is bijectively mapped to the pair of macro variables, energy density U and initial magnetization M_0, as

$$U = \frac{(\Delta p)^2}{6} + \frac{1 - M_0^2}{2}, \qquad M_0 = \frac{\sin\Delta\theta}{\Delta\theta}, \tag{5}$$

where $0 \le M_0 \le 1$ and $U \ge (1 - M_0^2)/2$. We consider (M_0, U) plane as the two-dimensional parameter space each of whose point identifies an initial waterbag distribution. Note that the parameter plane includes initial quantity M_0 as a control parameter which does not appear in equilibrium statistical mechanics, since we consider short time relaxation to QSSs, and QSSs remember initial conditions.

3 Statistical Theory in Quasi-Stationary States

Temporal evolution of a Hamiltonian system is governed by the Liouville equation, and the Liouville equation is reduced to the Boltzmann equation

$$\frac{\partial f}{\partial t} + \frac{\partial \epsilon}{\partial p}\frac{\partial f}{\partial \theta} - \frac{\partial \epsilon}{\partial \theta}\frac{\partial f}{\partial p} = C, \tag{6}$$

by the BBGKY hierarchy. Here $f(\theta, p, t)$ is one-body distribution, C represents a collisional term including two-body distribution function, and $\epsilon(\theta, p, t)$ is the one-body Hamiltonian

$$\epsilon(\theta, p, t) = \frac{p^2}{2} - M(t)\cos\theta \tag{7}$$

with magnetization

$$M(t) = \int_{-\pi}^{\pi} d\theta \int_{-\infty}^{\infty} dp \, \cos\theta \, f(\theta, p, t). \tag{8}$$

Note that the one-body Hamiltonian (7) differs from energy par rotator (2).

In the violent relaxation, we may omit the collisional term C and hence f is governed by the Vlasov equation

$$\frac{\partial f}{\partial t} + \frac{\partial \epsilon}{\partial p}\frac{\partial f}{\partial \theta} - \frac{\partial \epsilon}{\partial \theta}\frac{\partial f}{\partial p} = 0. \tag{9}$$

The Vlasov equation implies that $f(\theta(t), p(t), t)$ is a constant of time if $(\theta(t), p(t))$ is a solution of canonical equations of motion

$$\frac{d\theta}{dt} = \frac{\partial \epsilon}{\partial p} = p, \quad \frac{dp}{dt} = -\frac{\partial \epsilon}{\partial \theta} = -M(t)\sin\theta.$$

Moreover, the above canonical equations of motion imply that any area on μ-space must be preserved. These preserving features of the Vlasov equation is the key idea to construct a statistical theory in QSSs.

Let us review the construction of the statistical theory following Lynden-Bell (1967). We divide μ-space into mesoscopic cells, and we also divide each mesoscopic cell into microscopic cells. An initial distribution is set as waterbag (3), and we regard the value f_0 on a microscopic cell as a classical particle. Initially microscopic cells on the domain D have the same value of f_0, and hence the classical particles are *indistinguishable*. The particles move on μ-space following the Vlasov equation, but two particles cannot share one microscopic cell since the value of f must be 0 or f_0 on a microscopic cell due to the preserving features of the Vlasov equation. The particles are hence *exclusive*.

Counting the number of cases to distribute the indistinguishable and exclusive particles among microscopic cells, Lynden-Bell (1967) introduced the entropy S for a course-grained distribution $\bar{f}$ as

$$S[\bar{f}] = -\int_{-\pi}^{\pi} d\theta \int_{-\infty}^{\infty} dp \left[\frac{\bar{f}}{f_0} \log \frac{\bar{f}}{f_0} + \left(1 - \frac{\bar{f}}{f_0}\right) \log \left(1 - \frac{\bar{f}}{f_0}\right) \right], \tag{10}$$

where $\bar{f}(\theta, p, t)$ is proportional to number of particles in a mesoscopic cell including the point (θ, p). Hereafter we omit the bar of $\bar{f}$ for simplicity of the symbol.

Entropy (10) must be maximized with keeping constraints of conservation laws. The HMF model has three conservative quantities: probability which must be unity, energy density U and momentum density P. The conservation laws for the three quantities are written as

$$\begin{aligned} 1 &= \int_{-\pi}^{\pi} d\theta \int_{-\infty}^{\infty} dp \; f(\theta, p, t), \\ U &= \int_{-\pi}^{\pi} d\theta \int_{-\infty}^{\infty} dp \left(\frac{p^2}{2} + \frac{1 - M(t)\cos\theta}{2} \right) f(\theta, p, t), \end{aligned} \tag{11}$$

and

$$P = \int_{-\pi}^{\pi} d\theta \int_{-\infty}^{\infty} dp \; p f(\theta, p, t).$$

We remark that the right-hand-side of the second equation does not consist of the one-body Hamiltonian (7), but consist of energy par rotator (2). Using the Vlasov equation (9) and integration by parts, one can show that U is a constant of time thanks to the factor $1/2$ of the potential term.

Maximizing entropy (10) with these constraints, we obtain distribution for QSS as

$$f(\theta, p) = \frac{f_0}{\exp(\alpha + \beta(p^2/2 - M\cos\theta) + \lambda p) + 1}, \tag{12}$$

where α, β and λ are the Lagrange's multipliers for probability, energy density and momentum density respectively. From the symmetry of initial distribution with respect to p, momentum density P must be zero, and hence $\lambda = 0$. The obtained distribution is a stationary solution of the Vlasov equation and hence we omit the argument t. We still have three unknown parameters: α, β and magnetization M. Values of the three parameters are determined by substituting the distribution (12) into the three equations: the two conservation laws (11) and the definition of magnetization (8). Note that the last equation is a self-consistent equation for M. We denote a solution of the self-consistent equation by $M_{\rm QSS}$ to clearly indicate that the solution is for QSS.

Summarizing, the distribution for QSS is given by (12), and the three parameters α, β and M are determined by solving a set of three equations (11) and (8). The three equations solely depend on initially given values of (f_0, U) which is equivalent to (M_0, U) through (5). The distribution (12) is hence obtained for a given initial waterbag distribution without any free parameters.

4 Numerical Results

In this section we numerically confirm validity of the theoretically predicted distribution (12). The theory assumes that temporal evolution of a state is described by the Vlasov equation up to QSS, then we confirm that this assumption holds first. After that, we compare the theoretically computed critical curve with numerical one, where the curve divides the two-dimensional parameter space (M_0, U) into magnetized phase $M_{\rm QSS} > 0$ and non-magnetized phase $M_{\rm QSS} = 0$.

All numerical computations are performed by integrating canonical equations of motion for the HMF model (1) by using fourth order symplectic integrator (Yoshida, 1993) with the fixed time slice $\Delta t = 0.2$. Initial values of $\theta_j(0)$ and $p_j(0)$ are randomly picked up from a waterbag distribution expressed by (3) and (4). Total momentum is set as 0.

Let us start to check the validity of the Vlasov description. The Vlasov equation (9) is obtained by omitting the collisional term C of the Boltzmann equation (6). The collisional term C is derived from the BBGKY hierarchy, and hence it depends on N, while the Vlasov equation does not depend on N. We can therefore conclude that temporal evolution of the system is described by the Vlasov equation, if no dependence on N appears.

We observe temporal evolution of the system through magnetization, which is reported in Fig. 1 for $N = 10^3$, 10^4, 10^5 and 10^6 with $U = 0.55$ and $M_0 = 0$. Due to finiteness, the initial magnetization is not exactly 0, but is typically read as $M_0 \sim N^{-1/2}$, since initial values of θ_j is randomly picked up from a waterbag distribution. The N dependence of M_0 prevents the four curves reported in Fig. 1a from coinciding with each other. However, magnetization exponentially growth up to the QSS level (Yamaguchi et al. 2004), which is indicated as the horizontal line in each panel of Fig. 1, and hence the N dependence can be removed by shifting the origin of time. The shifted curves are reported in Fig. 1b, and the four curves almost collapse in the early time region $t \lesssim 60$. The collapsing implies that the system is approximately governed by the Vlasov equation in this early time region.

Let us progress to compare the theoretical prediction of critical curve with numerically obtained one. The theoretical prediction is shown in Fig. 2d, and upper and lower regions represent non-magnetized phase $M_{\rm QSS} = 0$ and magnetized phase $M_{\rm QSS} > 0$ respectively. Note that $M_{\rm QSS} = 0$ is always a solution of the

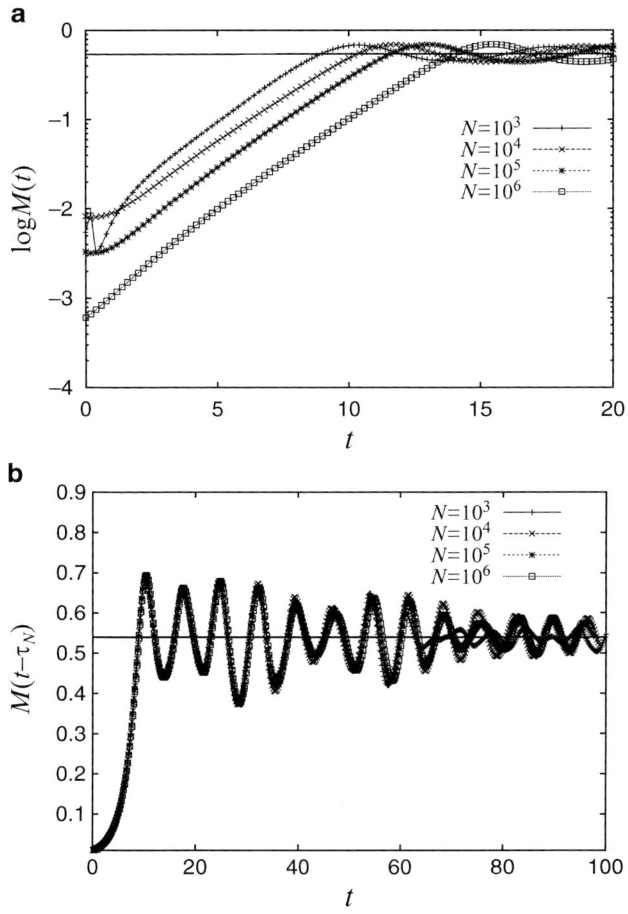

Fig. 1 Temporal evolution of magnetization M for $N = 10^3,\ 10^4,\ 10^5$ and 10^6. (**a**) Log-linear plot of $M(t)$. Lower curve corresponds to larger N. (**b**) Four curves are reported with time shifts τ_N, and they almost collapse in $t \lesssim 60$

self-consistent equation (8). The magnetized phase means that a non-zero solution also exists, and entropy at the non-zero solution is larger than at the zero solution.

We remark the two limits $M_0 \to 1$ and $M_0 \to 0$. In the former limit the critical curve goes to $U = 3/4$, and the value is the critical energy of second order phase transition in thermal equilibrium. In the latter limit the curve converges to $U = 7/12$, which corresponds to the destabilization of the non-magnetized initial state in the Vlasov equation (Yamaguchi et al. 2004).

We compare the theoretical prediction with numerically obtained density plots of M_{QSS} reported in Figs. 2a, b and c for $N = 10^3,\ 10^4$ and 10^5 respectively. In these three panels, a point becomes brighter as M_{QSS} takes a larger value. That is, the upper dark region corresponds to the non-magnetized phase, and the lower bright

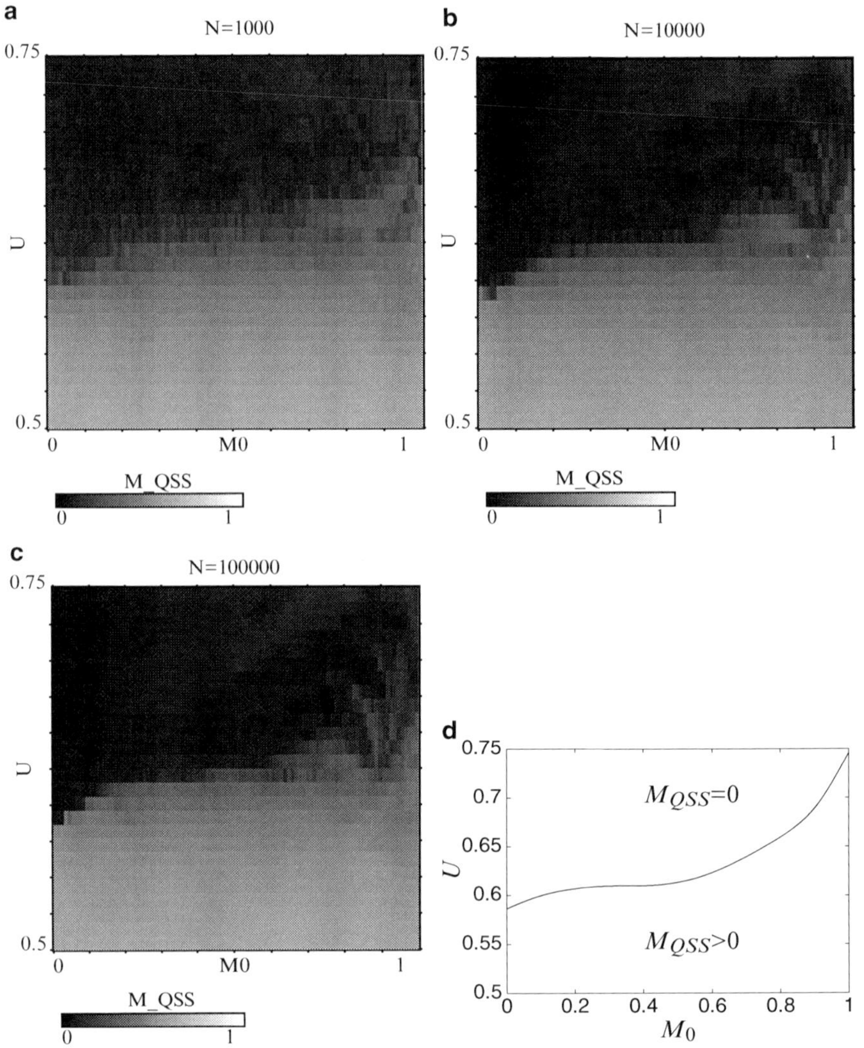

Fig. 2 Density plots of $M_{\rm QSS}$ for $N = 10^3$ (**a**) , 10^4 (**b**) and 10^5 (**c**) with the theoretically obtained critical curve (**d**). Dark and bright points correspond magnetized ($M_{\rm QSS} > 0$) and non-magnetized ($M_{\rm QSS} = 0$) phases respectively. The values of $M_{\rm QSS}$ are calculated as time averages of one orbit in the period $20 < t \leq 100$

region to the magnetized phase. The boundary between the two regions represents the critical curve, and it becomes clearer as N increases. Moreover, the numerically obtained critical curve is in good agreement with the theoretical prediction, particularly in a small M_0 region.

The numerically obtained boundary loses smoothness as M_0 increases, and hence the theoretical prediction becomes worse. Let us consider a typical case, $M_0 = 1$.

The initial condition $M_0 = 1$ is expressed as a one-dimensional interval on p-axis, and hence a particle on μ-space, which is a microscopic cell with $f = f_0$ introduced in Sec. 3, is not surrounded by other particles except for two located up and down. As a consequence, exclusivity, which is a key mechanics to derive the statistical theory, does not work well, and hence the statistical theory is not accurate.

5 Summary

We introduced a statistical theory for quasi-stationary states, which is based on the Vlasov description of a Hamiltonian system with long-range interactions. The statistical theory predicts a one-body distribution in QSS for a given initial waterbag distribution without any free parameters.

In the HMF model, we confirmed that the Vlasov description is valid in an early time region by showing N independence of temporal evolutions of magnetization. Moreover, the theoretically predicted critical curve is in good agreement with numerical simulations. The agreement becomes clearer as number of particles increases.

In this article we considered a family of simple waterbag initial distributions in the HMF model, but philosophy of the statistical theory is applicable to other types of initial distributions and models. Such extensions are interesting future works.

Acknowledgements The author thanks Andorea Antoniazzi, Duccio Fanelli and Stefano Ruffo for fruitful discussions. The author has been supported by the Ministry of Education, Science, Sports and Culture, Grant-in-Aid for Young Scientists (B), 19760052, 2007.

References

Antoni, M., Ruffo, S.: Phys. Rev. E **52**: 2361 (1995).
Antoniazzi, A., Fanelli, D., Barré, J., Chavanis, P.-H., Dauxois, T., Ruffo, S.: Phys. Rev. E **75**: 011112 (2007).
Antoniazzi, A., Fanelli, D., Ruffo, S., Yamaguchi, Y.Y.: Physica A **99**: 040601 (2007).
Balescu, R.: Statistical Dynamics – Matter Out of Equilibrium. Imperial College Press, Belgium (1997).
Barré, J., Bouchet, F., Dauxois, T., Ruffo, S., Yamaguchi, Y.Y.: Physica A **365**: 177 (2006).
Dauxois, T. et al. (eds): Dynamics and Thermodynamics of Systems with Long-Range Interactions. Lect. Notes Phys. Vol. 602, Springer, New York (2002).
Elskens, Y., Escande, D.F.: Microscopic Dynamics of Plasmas and Chaos. Institute of Physics, Bristol (2003).
Inagaki, S.: Prog. Theor. Phys. **90**: 577 (1993).
Lynden-Bell, D.: MNRAS **136**: 101 (1967).
Tsuchiya, T., Konishi, T., Gouda, N.: Phys. Rev. E **50**: 2607 (1994).
Yamaguchi, Y.Y., Barré, J., Bouchet, F., Dauxois, T., Ruffo, S.: Physica A **337**: 36 (2004).
Yoshida, H.: Celestial Mech. Dyn. Astron. **56**: 27 (1993).

An Alternative Method for Solving a Certain Class of Fractional Kinetic Equations

R.K. Saxena, A.M. Mathai, and H.J. Haubold

Abstract An alternative method for solving the fractional kinetic equations solved earlier by Haubold and Mathai (Astrophysics and Space Science 273:53–63, 2000) and Saxena et al. (Astrophysics and Space Science 282:281–287, 2002; Physica A 344:653–664, 2004a; Astrophysics and Space Science 290:299–310, 2004b) is recently given by Saxena and Kalla (Applied Mathematics and Computation 199:504–511, 2007). This method can also be applied in solving more general fractional kinetic equations than the ones solved by the aforesaid authors. In view of the usefulness and importance of the kinetic equation in certain physical problems governing reaction-diffusion in complex systems and anomalous diffusion, the authors present an alternative simple method for deriving the solution of the generalized forms of the fractional kinetic equations solved by the aforesaid authors and Nonnenmacher and Metzler (Fractals 3:557–566, 1995). The method depends on the use of the Riemann-Liouville fractional calculus operators. It has been shown by the application of Riemann-Liouville fractional integral operator and its interesting properties, that the solution of the given fractional kinetic equation can be obtained in a straight-forward manner. This method does not make use of the Laplace transform.

Keywords Mittag-Leffler function · Riemann-Liouville fractional calculus operator · Differential equation · Laplce transform · Fractional kinetic equation

1 Introduction

The paper deals with the essential problem related to applications of Mittag-Leffler function and Riemann-Liouville fractional calculus operators to fractional order kinetic equations arising in modeling physical phenomena, governing diffusion in porous media and relaxation processes. As such it reveals the important role of these tools in applications of fractional calculus. The results are interesting and useful for wide range of applied scientists dealing with fractional order differential and

R.K. Saxena (✉), A.M. Mathai, and H.J. Haubold
Department of Mathematics and Statistics, Jai Narain Vyas University, Jodhpur-342 004, India

H.J. Haubold and A.M. Mathai (eds.), *Proceedings of the Third UN/ESA/NASA Workshop on the International Heliophysical Year 2007 and Basic Space Science*, Astrophysics and Space Science Proceedings, DOI 10.1007/978-3-642-03325-4_5,

fractional order integral equations. In a series of papers the authors have demonstrated the use of integral transforms in the solution of certain fractional kinetic equations (2002, 2004a, 2004b), reaction-diffusion equations (2006a, 2006b), and fractional differential equations governing nonlinear waves (2006c, 2006d). In the present paper it is shown by the application of Riemann-Liouville fractional calculus operators and its interesting properties that the given fractional kinetic equations can be easily solved. Fractional kinetic equations are studied by Zaslavsky (1994), Saichev and Zaslavsky (1997), Gloeckle and Nonnenmacher (1991), and Saxena et al. (2002, 2004a, 2004b) due to their importance in the solution of certain applied problems governing reaction and relaxation in complex systems and anomalous diffusion. The use of fractional kinetic equations in many problems arising in science and engineering can be found in the monographs by Podlubny (1999), Hilfer (2000), and Kilbas et al. (2006) and the various papers given therein. The Mittag-Leffler functions naturally occur as a solution of fractional order differential equation or a fractional order integral equation. Mittag-Leffler (1903) defined this function, known as Mittag-Leffler function in the literature, in terms of the power series

$$E_\alpha(z) := \sum_{k=0}^{\infty} \frac{z^k}{\Gamma(\alpha k + 1)}; \quad (\alpha \in C, Re(\alpha) > 0). \tag{1}$$

This function is generalized by Wiman (1905) in the form

$$E_{\alpha,\beta}(z) := \sum_{k=0}^{\infty} \frac{z^k}{\Gamma(\alpha k + \beta)}, (\alpha, \beta \in C, Re(\alpha) > 0, Re(\beta) > 0). \tag{2}$$

According to Dzherbashyan (1966, p.118), both the functions defined by (1) and (2) are entire functions of order $\rho = 1/\alpha$ and type $\sigma = 1$. A comprehensive detailed account of these functions is available from the monographs of Erdélyi, Magnus, Oberhettinger and Tricomi (1995, Chapt. 18) and Dzherbashyan (1966, Chapt. 2). The Riemann-Liouville operators of fractional calculus are defined in the books by Miller and Ross (1993), Oldham and Spanier (1974), Podlubny (1999) and Kilbas et al. (2006) as

$${}_\alpha D_t^{-\nu} N(t) := \frac{1}{\Gamma(\nu)} \int_\alpha^t (t-u)^{\nu-1} \; N(u)du, Re(\nu) > 0, t > a \tag{3}$$

with ${}_a D_t^0 N(t) = N(t)$, and

$${}_a D_t^{\mu} N(t) := \frac{d^n}{dt^n}({}_a D_t^{\mu-n} N(t)), Re(\mu) > 0, n - \mu > 0. \tag{4}$$

By virtue of the definitions (3), it is not difficult to show that

$${}_a D_t^{-\nu} (t-a)^{\rho-1} = \frac{\Gamma(\rho)}{\Gamma(\rho+\nu)} (t-a)^{\rho+\nu-1}, \tag{5}$$

where $Re(\nu) > 0$, $Re(\rho) > 0; t > a$. Also from (Podlubny 1999, p.72, (2.117)), we have

$$ {}_aD_t^{\nu}(t-a)^{\rho-1} = \frac{\Gamma(\rho)}{\Gamma(\rho-\nu)}(t-a)^{\rho-\nu-1}, \tag{6} $$

where $Re(\nu) > 0$, $Re(\rho) > 0, t > a$. When $\rho = 1$ (6) reduces to an interesting formula

$$ {}_aD_t^{\nu}1 = \frac{1}{\Gamma(1-\nu)}(t-a)^{-\nu}, t > a; \nu \neq 1, 2 \ldots \tag{7} $$

which is a remarkable result in the theory of fractional calculus and indicates that the fractional derivative of a constant is not zero.

We now proceed to derive and solve the fractional kinetic equations in the next section.

2 Derivation of the Fractional Kinetic Equation and Its Solution

If we integrate the standard kinetic equation

$$ \frac{d}{dt}N_i(t) = -c_i N_i(t), (c_i > 0) \tag{8} $$

we obtain (Haubold and Mathai, 2000, p.58)

$$ N(t) - N_a = -c_i \; {}_aD_t^{-1}N_i(t), \tag{9} $$

where ${}_aD_t^{-1}$ is the standard Riemann integral operator. Here we recall that in the original paper of Haubold and Mathai (2000), the number density of species, $N_i = N_i(t)$ is a function of time. Further we assume that $N_i(t = a) = N_a$ is the number density of species i at time $t = a$. If we drop the index i in (9) and generalize it, we arrive at the fractional kinetic equation

$$ N(t) - N_a = -c^{\nu} \; {}_aD_t^{-\nu}N(t) \tag{10} $$

Solution of (10). If we multiply both sides of (10) by $(-c^{\nu})^m \; {}_aD_t^{-m\nu}$, we obtain

$$ (-c^{\nu})^m \; {}_aD_t^{-m\nu}N(t) - (-c^{\nu})(-c^{\nu})^m \; {}_aD_t^{-m\nu-\nu} \; N(t) = (-c^{\nu})^m \; {}_aD_t^{-m\nu}N_a. \tag{11} $$

Now summing up both sides of (11) for m from 0 to ∞, it yields

$$ \sum_{m=0}^{\infty}(-c^{\nu})^m \; {}_aD_t^{-m\nu}N(t) - \sum_{m=0}^{\infty}(-c^{\nu})^{m+1} \; {}_aD_t^{-m\nu-\nu}N(t) $$

$$ = N_a \sum_{m=0}^{\infty}(-c^{\nu})^m \; {}_aD_t^{-m\nu}1, \tag{12} $$

which on using the formula (5) yields

$$N(t) = N_a \sum_{m=0}^{\infty} (-c^\nu)^m [(t-a)^{m\nu} / \Gamma(m\nu + 1)] \tag{13}$$

$$= N_a E_\nu[-c^\nu (t-a)^\nu], t > a \tag{14}$$

Thus we arrive at the following theorem:

Theorem 2.1. *If* $Re(\nu) > 0$, $Re(c) > 0$ *then there exists the unique solution of the integral equation*

$$N(t) - N_a = -c^\nu \ {}_aD_t^{-\nu}(t), \tag{15}$$

given by

$$N(t) = N_a E_\nu(-c^\nu (t-a)^\nu), \ t > a \tag{16}$$

with the Mittag-Leffler function defined by (1).

When $a \to 0$, *(16) reduces to the following result given by Haubold and Mathai (2000, p.63)*:

Corollary 2.1. *If,* $Re(c) > 0$ *then the unique solution of the integral equation*

$$N(t) - N_0 = -c^\nu \ {}_0D_t^{-\nu}(t), \tag{17}$$

is given by

$$N(t) = N_0 E_\nu(-c^\nu t^\nu). \tag{18}$$

Remark 2.1. If we apply the operator ${}_aD_t^\nu$ from the left to (10) and make use of (7), we obtain the fractional differential equation

$${}_aD_t^\nu N(t) - N_a \frac{(t-a)^{-\nu}}{\Gamma(1-\nu)} = -c^\nu N(t), \ \ t > a \tag{19}$$

whose solution is also given by (16). When a tends to zero in (16), it reduces to one obtained by Nonnenmacher and Metzler (1995, p.156) for the fractional relaxation equation with c replaced by $1/c$.

Remark 2.2. The method adopted in deriving the solution of fractional kinetic equation (8) is similar to that used by Al-Saqabi and Tuan (1996) for solving differ integral equations.

Theorem 2.2. *If* $min\{Re(\nu), Re(\mu)\} > 0$, $Re(c) > 0$, *then there exists the unique solution of the integral equation*

$$`N(t) - N_a t^{\mu-1} = -c^\nu \ {}_aD_t^{-\nu}(t), \tag{20}$$

given by

$$N(t) = N_a \Gamma(\mu)(t-a)^{\mu-1} \ E_{\nu,\mu}(-c^\nu (t-a)^\nu), t > a \tag{21}$$

where $E_{\nu,\mu}(t)$*is the generalized Mittag-Leffler function defined by (2).*

Solution of (20). If we multiply both sides of (20) by $(-c^{\nu})^m \ {}_aD_t^{-m\nu}$, we obtain

$$(-c^{\nu})^m \ {}_aD_t^{-m\nu}N(t) - (-c^{\nu})(-c^{\nu})^m \ {}_aD_t^{-m\nu-\nu}N(t) = N_a(-c^{\nu})^m \ {}_aD_t^{-m\nu}t^{\mu-1}. \tag{22}$$

Now summing up both sides of (22) for m from 0 to ∞, it yields

$$\sum_{m=0}^{\infty}(-c^{\nu})^m \ {}_aD_t^{-m\nu}N(t) - \sum_{m=0}^{\infty}(-c^{\nu})^{m+1} \ {}_aD_t^{-m\nu-\nu}N(t)$$
$$= N_a \sum_{m=0}^{\infty}(-c^{\nu})^m \ {}_aD_t^{-m\nu}t^{\mu-1}, \tag{23}$$

which on using the formula (5) gives

$$N(t) = N_a\Gamma(\mu)\sum_{m=0}^{\infty}(-c^{\nu})^m[(t-a)^{m\nu}/\Gamma(m\nu+\mu)] \tag{24}$$
$$= N_a E_{\nu,\mu}[-c^{\nu}(t-a)^{\nu}], \ t > a. \tag{25}$$

This completes the proof of Theorem 2.2.

For $a = 0$, (25) reduces to the following result given by Saxena et al. (2002, p. 283).

Corollary 2.2. *If min* $\{Re(\nu), Re(\mu)\} > 0$, $R(c) > 0$ *then the solution of the integral equation*

$$N(t) - N_0t^{\mu-1} = -c^{\nu} \ {}_0D_t^{-\nu}(t) \tag{26}$$

' is given by

$$N(t) = N_0\Gamma(\mu)t^{\mu-1} \ E_{\nu,\mu}(-c^{\nu}t^{\nu}), \tag{27}$$

where $E_{\nu,\mu}(t)$ *is the generalized Mittag-Leffler function defined by (2).*

References

Al-Saqabi, B.N., Tuan, V.K.: Solution of a fractional differ integral equation. Integral Transforms and Special Functions **4**, 321–326 (1996)

Dzherbashyan, M.M.: Integral Transforms and Representation of Functions in Complex Domain (in Russian). Nauka, Moscow (1966)

Dzherbashyan, M.M.: Harmonic Analysis and Boundary Value Problems in the Complex Domain. Birkhaeuser-Verlag, Basel and London (1993)

Erdélyi, A., Magnus, W., Oberhettinger, F., Tricomi, F.G.: Tables of Integral Transforms. Vol. **2**, McGraw-Hill, New York, Toronto, and London (1954)

Erdélyi, A., Magnus, W., Oberhettinger, F., Tricomi, F.G.: Higher Transcendental Functions. Vol. **3**, McGraw-Hill, New York, Toronto, and London (1955)

Gloeckle, W.G., Nonnenmacher, T.F.: Fractional integral operators and Fox function in the theory of viscoelasticity. Macromolecules **24**, 6426–6434 (1991)

Haubold, H.J., Mathai, A.M.: The fractional kinetic equation and thermonuclear functions. Astrophysics and Space Science **273**, 53–63 (2000)
Hilfer, R.: Applications of Fractional Calculus in Physics. World Scientific Publishing Co., New York (2000)
Kilbas, A.A., Srivastava, H.M., Trujillo, J.J.: Theory and Applications of Fractional Differential Equations. North-Holland Mathematics Studies **204**, Elsevier, Amsterdam (2006)
Mathai, A.M., Saxena, R.K.: The H-function with Applications in Statistics and Other Disciplines. John Wiley and Sons Inc., New York, London and Sydney (1978)
Miller, K.S., Ross, B.: An Introduction to Fractional Calculus and Fractional Differential Equations. Wiley and Sons, New York (1993)
Mittag-Leffler, G.M.: Sur la nouvelle function. C.R. Acad. Sci., Paris, **137**, 554–558 (1903)
Nonnenmacher, T.F., Metzler, R.: On the Riemann-Liouville fractional calculus and some recent applications. Fractals **3**, 557–566 (1995)
Oldham, K.B., Spanier, J.: The Fractional Calculus: Theory and Applications of Differentiation and Integration of Arbitrary Order. Academic Press, New York (1974)
Podlubny, I.: Fractional Differential Equations. Academic Press, San Diego (1999)
Saichev, A., Zaslavsky, M.: Fractional kinetic equations: solutions and applications, Chaos **7**, 753–764 (1997)
Saxena, R.K., Kalla, S.L.: 2007, On the solution of certain fractional kinetic equations. In Applied Mathematics and Computation **199**, 504–511 (2007)
Saxena, R.K., Mathai, A.M., Haubold, H.J.: On fractional kinetic equations. Astrophysics and Space Science **282**, 281–287 (2002)
Saxena, R.K., Mathai, A.M., Haubold, H.J.: On generalized fractional kinetic equations. Physica A **344**, 653–664 (2004a)
Saxena, R.K., Mathai, A.M., Haubold, H.J.: Unified fractional kinetic equation and a fractional diffusion equation. Astrophysics and Space Science **290**, 299–310 (2004b)
Saxena, R.K., Mathai, A.M., Haubold, H.J.: Fractional reaction-diffusion equations. Astrophysics and Space Science **305**, 289–296 (2006a)
Saxena, R.K., Mathai, A.M., Haubold, H.J.: Solution of generalized fractional reaction-diffusion equations. Astrophysics and Space Science **305**, 305–313 (2006b)
Saxena, R.K., Mathai, A.M., Haubold, H.J.: Reaction-diffusion systems and nonlinear waves. Astrophysics and Space Science **305**, 297–303 (2006c)
Saxena, R.K., Mathai, A.M., Haubold, H.J.: Solution of fractional reaction-diffusion equations in terms of Mittag-Leffler functions. International Journal of Scientific Research **15**, 1–17 (2006d)
Wiman, A.: Ueber den Fundamentalsatz in der Theorie der Funktionen. Acta Mathematica **29**, 191–201 (1905)
Zaslavsky, G.M.: Fractional kinetic equation for Hamiltonian chaos. Physica D **78**, 110–122 (1994)

Extended Reaction Rate Integral as Solutions of Some General Differential Equations

D.P. Joseph and H.J. Haubold

Abstract Here an extended form of the reaction rate probability integral, in the case of nonresonant thermonuclear reactions with the depleted tail and the right tail cut off, is considered. The reaction rate integral then can be looked upon as the inverse of the convolution of the Mellin transforms of Tsallis type statistics of nonextensive statistical mechanics and stretched exponential as well as that of superstatistics and stretched exponentials. The differential equations satisfied by the extended probability integrals are derived. The idea used is a novel one of evaluating the extended integrals in terms of some special functions and then by invoking the differential equations satisfied by these special functions. Some special cases of limiting situations are also discussed.

Keywords Reaction rate probability integrals · Extended probability integrals · Pathway model · H-function · G-function · Differential equations

1 Introduction

Nuclear reactions govern major aspects of the chemical evolution of the universe. A proper understanding of the nuclear reactions that are going on in hot cosmic plasma, and those in the laboratories as well, requires a sound theory of nuclear-reaction dynamics. The reaction probability integral is the probability per unit time that two particles, confined to a unit volume, will react with each other Haubold and Mathai (1998). In the present article we will show that one can obtain the reaction rate probability integral as a solution of a certain differential equation, the technique used is a novel one. First we evaluate the integral and represent it as a special function Mathai (1993); Luke (1969). Then we can invoke the differential equation satisfied by this special function, thereby establishing the differential equation for the reaction rate integral. A study of this integral as a solution of a differential

D.P. Joseph (✉) and H.J. Haubold
Centre for Mathematical Sciences, Pala Campus, Pala-686 574, Kerala, India

H.J. Haubold and A.M. Mathai (eds.), *Proceedings of the Third UN/ESA/NASA Workshop on the International Heliophysical Year 2007 and Basic Space Science*, Astrophysics and Space Science Proceedings, DOI 10.1007/978-3-642-03325-4_6,

equation is made because the behavior of physical systems is usually studied with the help of differential equations and hence the differential equations derived in the present paper may become useful one day.

Here we will consider the integrals of the following form:

Nonresonant case with depleted tail: (for details of the integrals see Haubold and Mathai, 1998)

$$I_1^{(\delta)} = \int_0^\infty x^{\alpha-1} e^{-ax^\delta - bx^{-\rho}} \mathrm{d}x, \;\; a>0,\; b>0,\; \delta>0,\; \rho>0. \tag{1}$$

Nonresonant case with depleted tail and high energy cut-off:

$$I_2^{(\delta)} = \int_0^d x^{\alpha-1} e^{-ax^\delta - bx^{-\rho}} \mathrm{d}x, \;\; a>0,\; b>0,\; \delta>0,\; \rho>0,\; d<\infty. \tag{2}$$

Do these integrals, as functions of b, satisfy some differential equations? No such differential equations can be easily seen from the integrals. We will show that these integrals will satisfy certain differential equations. The main results will be stated as follows. We will establish the following theorems for $I_1^{(\delta)}$ and extended $I_{1\beta}^{(\delta)}$ and $I_{2\beta}^{(\delta)}$.

Theorem 1.1. *A constant multiple of the reaction rate integral in (1), under the condition $\frac{\delta}{\rho} = m, m = 1, 2, \cdots$, can be obtained as a solution of the differential equation,*

$$\left[(-1)^{m+1} z - \eta\left(\eta - \frac{1}{m}\right)\cdots\left(\eta - \frac{m-1}{m}\right)\left(\eta - \frac{\alpha}{m\rho}\right)\right] f(z) = 0, \tag{3}$$

where, $z = \frac{ab^m}{m^m}$, $\eta = z\frac{\mathrm{d}}{\mathrm{d}z}$, $f(z) = \frac{\rho a^{\frac{\alpha}{m\rho}} m^{\frac{1}{2}}}{(2\pi)^{\frac{1-m}{2}}} I_1^{(m\rho)}(z)$ *and*

$$I_1^{(m\rho)}(z) = \frac{(2\pi)^{\frac{1-m}{2}}}{\rho a^{\frac{\alpha}{m\rho}} m^{\frac{1}{2}}} G^{m+1,0}_{0,m+1}\left[\frac{ab^m}{m^m}\Big|_{0,\frac{1}{m},\cdots,\frac{m-1}{m},\frac{\alpha}{m\rho}}\right], \;\; a>0,\; b>0,\; \alpha>0,\; \rho>0. \tag{4}$$

Corollary 1.1. *When $\delta = 1$, a constant multiple of the reaction rate integral in (1), under the condition $\frac{1}{\rho} = m, m = 1, 2, \cdots$, can be obtained as a solution of the differential equation,*

$$\left[(-1)^{m+1} z - \eta\left(\eta - \frac{1}{m}\right)\cdots\left(\eta - \frac{m-1}{m}\right)(\eta - \alpha)\right] f(z) = 0, \tag{5}$$

where, $z = \frac{ab^m}{m^m}$, $\eta = z\frac{\mathrm{d}}{\mathrm{d}z}$, $f(z) = \frac{\rho a^{\alpha} m^{\frac{1}{2}}}{(2\pi)^{\frac{1-m}{2}}} I_1(z)$ *and*

$$I_1(z) = \frac{(2\pi)^{\frac{1-m}{2}}}{\rho a^{\alpha} m^{\frac{1}{2}}} G^{m+1,0}_{0,m+1}\left[\frac{ab^m}{m^m}\bigg|_{0,\frac{1}{m},\cdots,\frac{m-1}{m},\alpha}\right], \ a>0,\ b>0,\ \alpha>0,\ \rho>0. \tag{6}$$

The proof of the theorem will be given later.

Now we will consider a wider class of integrals as extended forms of (1) and (2). Consider the integrals

$$I^{(\delta)}_{1\beta} = \int_0^{\infty} x^{\alpha-1}[1+a(\beta-1)x^{\delta}]^{-\frac{1}{\beta-1}}\mathrm{e}^{-bx^{-\rho}}\mathrm{d}x,\ a>0,\ b>0,\ \delta>0,\ \rho>0,\ \beta>1, \tag{7}$$

$$I^{(\delta)}_{2\beta} = \int_0^{d} x^{\alpha-1}[1-a(1-\beta)x^{\delta}]^{\frac{1}{1-\beta}}\mathrm{e}^{-bx^{-\rho}}\mathrm{d}x,\ a>0,\ b>0,\ \delta>0,\ \rho>0,\ \beta<1. \tag{8}$$

These are obtained by replacing $\mathrm{e}^{-ax^{\delta}}$ *by* $[1-a(1-\beta)x^{\delta}]^{\frac{1}{1-\beta}}$ *so that when* $\beta\to 1$ *we have the integrals in (1) and (2). Thus* $I^{(\delta)}_{1\beta}$ *and* $I^{(\delta)}_{2\beta}$ *are the extended families of integrals in* $I^{(\delta)}_1$ *and* $I^{(\delta)}_2$ *respectively. Here we obtain the following main results for the extended forms of the reaction probability integrals.*

Theorem 1.2. *A constant multiple of the extended reaction rate integral in (7), under the condition* $\frac{\delta}{\rho} = m, m = 1,2,\cdots$, *can be obtained as a solution of the differential equation,*

$$\left[(-1)^{m+1}z\left(\eta+\frac{1}{\beta-1}-\frac{\alpha}{m\rho}\right)-\eta\left(\eta-\frac{1}{m}\right)\cdots\left(\eta-\frac{m-1}{m}\right)\left(\eta-\frac{\alpha}{m\rho}\right)\right]f(z)=0, \tag{9}$$

where, $z = \frac{a(\beta-1)b^m}{m^m}$, $\eta = z\frac{\mathrm{d}}{\mathrm{d}z}$, $f(z) = \frac{\rho[a(\beta-1)]^{\frac{\alpha}{m\rho}}\Gamma\left(\frac{1}{\beta-1}\right)m^{\frac{1}{2}}}{(2\pi)^{\frac{1-m}{2}}} I^{m\rho}_{1\beta}(z)$ *and*

$$I^{m\rho}_{1\beta}(z) = \frac{(2\pi)^{\frac{1-m}{2}}}{\rho[a(\beta-1)]^{\frac{\alpha}{m\rho}}\Gamma\left(\frac{1}{\beta-1}\right)m^{\frac{1}{2}}} G^{m+1,1}_{1,m+1}\left[\frac{a(\beta-1)b^m}{m^m}\bigg|^{\frac{\beta-2}{\beta-1}+\frac{\alpha}{m\rho}}_{0,\frac{1}{m},\cdots,\frac{m-1}{m},\frac{\alpha}{m\rho}}\right], \tag{10}$$

$a>0,\ b>0,\ \alpha>0,\ \rho>0,\ \beta>1.$

Corollary 1.2. *When* $\delta = 1$, *a constant multiple of the extended reaction rate integral in (7), under the condition* $\frac{1}{\rho} = m, m = 1,2,\cdots$, *can be obtained as a solution of the differential equation,*

$$\left[(-1)^{m+1}z\left(\eta+\frac{1}{\beta-1}-\alpha\right)-\eta\left(\eta-\frac{1}{m}\right)\cdots\left(\eta-\frac{m-1}{m}\right)(\eta-\alpha)\right]f(z)=0, \tag{11}$$

where, $z = \frac{a(\beta-1)b^m}{m^m}$, $\eta = z\frac{\mathrm{d}}{\mathrm{d}z}$, $f(z) = \frac{\rho[a(\beta-1)]^\alpha \Gamma\left(\frac{1}{\beta-1}\right) m^{\frac{1}{2}}}{(2\pi)^{\frac{1-m}{2}}} I_{1\beta}(z)$ *and*

$$I_{1\beta}(z) = \frac{(2\pi)^{\frac{1-m}{2}}}{\rho[a(\beta-1)]^\alpha \Gamma\left(\frac{1}{\beta-1}\right) m^{\frac{1}{2}}} G^{m+1,1}_{1,m+1}\left[\frac{a(\beta-1)b^m}{m^m}\Big|^{\frac{\beta-2}{\beta-1}+\alpha}_{0,\frac{1}{m},\cdots,\frac{m-1}{m},\alpha}\right], \tag{12}$$

$a > 0,\ b > 0,\ \alpha > 0,\ \rho > 0,\ \beta > 1.$

Theorem 1.3. *A constant multiple of the extended reaction rate integral in (8), under the condition* $\frac{\delta}{\rho} = m, m = 1, 2, \cdots$, *can be obtained as a solution of the differential equation,*

$$\left[(-1)^m z\left(\eta - \frac{1}{1-\beta} - \frac{\alpha}{m\rho}\right) - \eta\left(\eta - \frac{1}{m}\right)\cdots\left(\eta - \frac{m-1}{m}\right)\left(\eta - \frac{\alpha}{m\rho}\right)\right] f(z) = 0, \tag{13}$$

where, $z = \frac{a(1-\beta)b^m}{m^m}$, $\eta = z\frac{\mathrm{d}}{\mathrm{d}z}$, $f(z) = \frac{\rho[a(1-\beta)]^{\frac{\alpha}{m\rho}} m^{\frac{1}{2}}}{\Gamma\left(\frac{2-\beta}{1-\beta}\right)(2\pi)^{\frac{1-m}{2}}} I^{m\rho}_{2\beta}(z)$ *and*

$$I^{m\rho}_{2\beta}(z) = \frac{\Gamma\left(\frac{2-\beta}{1-\beta}\right)(2\pi)^{\frac{1-m}{2}}}{\rho[a(1-\beta)]^{\frac{\alpha}{m\rho}} m^{\frac{1}{2}}} G^{m+1,0}_{1,m+1}\left[\frac{a(1-\beta)b^m}{m^m}\Big|^{\frac{2-\beta}{1-\beta}+\frac{\alpha}{m\rho}}_{0,\frac{1}{m},\cdots,\frac{m-1}{m},\frac{\alpha}{m\rho}}\right], \tag{14}$$

$a > 0,\ b > 0,\ \alpha > 0,\ \rho > 0,\ \beta < 1.$

Corollary 1.3. *When* $\delta = 1$ *a constant multiple of the extended reaction rate integral in (8), under the condition* $\frac{1}{\rho} = m, m = 1, 2, \cdots$, *can be obtained as a solution of the differential equation,*

$$\left[(-1)^m z\left(\eta - \frac{1}{1-\beta} - \alpha\right) - \eta\left(\eta - \frac{1}{m}\right)\cdots\left(\eta - \frac{m-1}{m}\right)(\eta - \alpha)\right] f(z) = 0, \tag{15}$$

where, $z = \frac{a(1-\beta)b^m}{m^m}$, $\eta = z\frac{\mathrm{d}}{\mathrm{d}z}$, $f(z) = \frac{\rho[a(1-\beta)]^\alpha m^{\frac{1}{2}}}{\Gamma\left(\frac{2-\beta}{1-\beta}\right)(2\pi)^{\frac{1-m}{2}}} I_{2\beta}(z)$ *and*

$$I_{2\beta}(z) = \frac{\Gamma\left(\frac{2-\beta}{1-\beta}\right)(2\pi)^{\frac{1-m}{2}}}{\rho[a(1-\beta)]^\alpha m^{\frac{1}{2}}} G^{m+1,0}_{1,m+1}\left[\frac{a(1-\beta)b^m}{m^m}\Big|^{\frac{2-\beta}{1-\beta}+\alpha}_{0,\frac{1}{m},\cdots,\frac{m-1}{m},\alpha}\right], \tag{16}$$

$a > 0,\ b > 0,\ \alpha > 0,\ \rho > 0,\ \beta < 1.$
The proofs of the above theorems will be given after evaluating the integrals first.

2 Evaluation of the Integral $I_1^{(\delta)}$

The following procedure is available in Mathai and Haubold (1998). For the sake of completeness a brief outline of the derivation will be given here.

$$I_1^{(\delta)} = \int_0^{\infty} x^{\alpha-1} \mathrm{e}^{-\mathrm{a}x^{\delta}-\mathrm{b}x^{-\rho}} \mathrm{d}x, \quad \mathrm{a} > 0,\ \mathrm{b} > 0,\ \delta > 0,\ \rho > 0.$$

Here the integrand can be taken as a product of positive integrable functions and then we can apply statistical distribution theory to evaluate this integral. Let x_1 and x_2 be real scalar independent random variables having densities

$$f_1(x_1) = \begin{cases} c_1 x_1^{\alpha} \mathrm{e}^{-a x_1^{\delta}}, \ 0 < x_1 < \infty,\ a > 0,\ \delta > 0 \\ 0, \qquad \text{elsewhere} \end{cases} \tag{17}$$

and

$$f_2(x_2) = \begin{cases} c_2 \mathrm{e}^{-x_2^{\rho}}, \ 0 < x_2 < \infty,\ \rho > 0 \\ 0, \qquad \text{elsewhere,} \end{cases} \tag{18}$$

where c_1 and c_2 are normalizing constants. Let us transform x_1 and x_2 to $u = x_1 x_2$ and $v = x_1$. Then the marginal density of u is given by

$$g_1(u) = \int_v \frac{1}{v} f_1(v) f_2\left(\frac{u}{v}\right) \mathrm{d}v \tag{19}$$

$$= c_1 c_2 \int_0^{\infty} v^{\alpha-1} \mathrm{e}^{-\mathrm{a}v^{\delta}-\mathrm{b}v^{-\rho}} \mathrm{d}v, \ \text{where } \mathrm{b} = \mathrm{u}^{\rho},\ \delta > 0,\ \rho > 0. \tag{20}$$

Let us evaluate the density through expected values or moments.

$$E(u^{s-1}) = E(x_1^{s-1}) E(x_2^{s-1}) \tag{21}$$

due to statistical independence of x_1 and x_2.

$$E(x_1^{s-1}) = c_1 \int_0^{\infty} x_1^{\alpha+s-1} \mathrm{e}^{-a x_1^{\delta}} \mathrm{d}x_1.$$

Putting $y = a x_1^{\delta}$ and evaluating the integral as a gamma integral, one has,

$$E(x_1^{s-1}) = \frac{c_1}{\delta a^{\frac{\alpha}{\delta}+\frac{s}{\delta}}} \Gamma\left(\frac{\alpha}{\delta} + \frac{s}{\delta}\right), \quad \Re(\alpha + s) > 0, \tag{22}$$

where $\Re(.)$ denotes the real part of (.).

$$E(x_2^{s-1}) = c_2 \int_0^{\infty} x_2^{s-1} \mathrm{e}^{-x_2^{\rho}} \mathrm{d}x_2.$$

Putting $y = x_2^{\rho}$, we get

$$E(x_2^{s-1}) = \frac{c_2}{\rho}\Gamma\left(\frac{s}{\rho}\right), \quad \Re(s) > 0. \tag{23}$$

From (22) to (23)

$$E(u^{s-1}) = \frac{c_1 c_2}{\delta\rho a^{\frac{\alpha}{\delta}+\frac{s}{\delta}}}\Gamma\left(\frac{s}{\rho}\right)\Gamma\left(\frac{\alpha}{\delta}+\frac{s}{\delta}\right), \; \Re(s) > 0, \; \Re(\alpha + s) > 0. \tag{24}$$

Looking at the $(s-1)^{\text{th}}$ moment as the Mellin transform of the corresponding density and then taking the inverse Mellin transform we get the density of u,

$$g_1(u) = \frac{c_1 c_2}{\delta\rho a^{\frac{\alpha}{\delta}}}\frac{1}{2\pi i}\int_{c-i\infty}^{c+i\infty}\Gamma\left(\frac{s}{\rho}\right)\Gamma\left(\frac{\alpha}{\delta}+\frac{s}{\delta}\right)\left(a^{\frac{1}{\delta}}b^{\frac{1}{\rho}}\right)^{-s}\mathrm{d}s. \tag{25}$$

Comparing (20) and (25)

$$\begin{aligned} I_1^{(\delta)} &= \int_0^{\infty} x^{\alpha-1}e^{-ax^{\delta}-bx^{-\rho}}\mathrm{d}x \\ &= \frac{1}{\delta\rho a^{\frac{\alpha}{\delta}}}\frac{1}{2\pi i}\int_{c-i\infty}^{c+i\infty}\Gamma\left(\frac{s}{\rho}\right)\Gamma\left(\frac{\alpha}{\delta}+\frac{s}{\delta}\right)\left(a^{\frac{1}{\delta}}b^{\frac{1}{\rho}}\right)^{-s}\mathrm{d}s. \end{aligned} \tag{26}$$

This contour integral can be written as an H-function Mathai and Saxena (1978). That is,

$$I_1^{(\delta)} = \frac{1}{\delta\rho a^{\frac{\alpha}{\delta}}}H_{0,2}^{2,0}\left[a^{\frac{1}{\delta}}b^{\frac{1}{\rho}}\Big|_{(0,\frac{1}{\rho}),(\frac{\alpha}{\delta},\frac{1}{\delta})}\right], \; a > 0, \; b > 0, \; \alpha > 0, \; \delta > 0, \; \rho > 0. \tag{27}$$

Make the transformation $\frac{s}{\delta} = s_1$ in (26)

$$I_1^{(\delta)} = \frac{1}{\rho a^{\frac{\alpha}{\delta}}}\frac{1}{2\pi i}\int_{c-i\infty}^{c+i\infty}\Gamma\left(\frac{\delta s_1}{\rho}\right)\Gamma\left(\frac{\alpha}{\delta}+s_1\right)\left(ab^{\frac{\delta}{\rho}}\right)^{-s_1}\mathrm{d}s_1. \tag{28}$$

Let us consider the special case where $\frac{\delta}{\rho} = m, \; m = 1, 2, \cdots$. In physics problems $\rho = \frac{1}{2}$ and δ an integer. Then this assumption of $\frac{\delta}{\rho} = m, \; m = 1, 2, \cdots$ is meaningful at least in some physical problems. That is,

$$I_1^{(m\rho)} = \frac{1}{\rho a^{\frac{\alpha}{m\rho}}}\frac{1}{2\pi i}\int_{c_1-i\infty}^{c_1+i\infty}\Gamma(ms_1)\Gamma\left(\frac{\alpha}{m\rho}+s_1\right)(ab^m)^{-s_1}\mathrm{d}s_1.$$

This can be reduced to a G-function by using the multiplication formula for gamma functions, namely,

$$\Gamma(mz) = (2\pi)^{\frac{1-m}{2}} m^{mz-\frac{1}{2}} \Gamma(z)\Gamma\big(z+\frac{1}{m}\big)\cdots\Gamma\big(z+\frac{m-1}{m}\big),\ m=1,2,\cdots. \quad (29)$$

Then we have,

$$\begin{aligned} I_1^{(m\rho)} &= \frac{(2\pi)^{\frac{1-m}{2}}}{\rho a^{\frac{\alpha}{m\rho}} m^{\frac{1}{2}}} \frac{1}{2\pi i} \int_{c_1-i\infty}^{c_1+i\infty} m^{ms_1} \Gamma(s_1)\Gamma\big(s_1+\frac{1}{m}\big)\cdots\Gamma\big(s_1+\frac{m-1}{m}\big) \\ &\quad \Gamma\big(s_1+\frac{\alpha}{m\rho}\big)(ab^m)^{-s_1}\mathrm{d}s_1. \end{aligned} \quad (30)$$

On evaluating (30), we get (4). (see Mathai and Haubold (1998); Saxena (1960); Haubold and Mathai (1984).

Particular Case, $\delta = 1$

When $\delta = 1$, we get the reaction rate integral in (1), under the condition $\frac{1}{\rho} = m$, $m = 1, 2, \cdots$ as,

$$I_1 = \int_0^\infty x^{\alpha-1} e^{-ax-bx^{-\rho}}\mathrm{d}x,\ a>0,\ b>0,\ \rho>0. \quad (31)$$

On evaluating this integral, we get (6). Observe that $m = 2$ is a real physical situation, see for example, Mathai and Haubold (1998).

3 Evaluation of Extended Integrals

We have,

$$I_{1\beta}^{(\delta)} = \int_0^\infty x^{\alpha-1}[1+a(\beta-1)x^\delta]^{-\frac{1}{\beta-1}} \mathrm{e}^{-bx^{-\rho}}\mathrm{d}x,\ a>0,\ b>0,\ \delta>0,\ \rho>0,\ \beta>1.$$

As $\beta \to 1$, $[1-a(1-\beta)x^\delta]^{\frac{1}{1-\beta}}$ becomes e^{-ax^δ} so that we can extend the reaction rate integrals in (1) and (2) using the pathway parameter β. Then here arise two cases:(i) $\beta < 1$, (ii) $\beta > 1$.

Case 3.1. $\beta < 1$

$$I_{2\beta}^{(\delta)} = \int_0^d x^{\alpha-1}[1-a(1-\beta)x^\delta]^{\frac{1}{1-\beta}} \mathrm{e}^{-bx^{-\rho}}\mathrm{d}x,\ a>0,\ b>0,\ \delta>0,\ \beta<1.$$

Here let us take

$$f_1(x_1) = \begin{cases} c_1 x_1^\alpha [1-a(1-\beta){x_1}^\delta]^{\frac{1}{1-\beta}},\ 0<x_1<\big[\frac{1}{a(1-\beta)}\big]^{\frac{1}{\delta}},\ a>0,\ \delta>0,\ \beta<1 \\ 0, \qquad \text{elsewhere} \end{cases} \quad (32)$$

and

$$f_2(x_2) = \begin{cases} c_2 \mathrm{e}^{-x_2^{\rho}}, \; 0 < x_2 < \infty, \; \rho > 0 \\ 0, \qquad \text{elsewhere,} \end{cases} \tag{33}$$

where c_1 and c_2 are the normalizing constants. Let us consider the case where $d = \left[\frac{1}{a(1-\beta)}\right]^{\frac{1}{\delta}}$ in (8). Note that $f_1(x_1)$ in (32) with $\alpha = 0$ and $\delta = 1$ is Tsallis statistics leading to nonextensive statistical mechanics. Also observe that the functional part of (32) for $\alpha = 0$ and $\delta = 1$ gives the power law as well.

$$\frac{\mathrm{d}}{\mathrm{d}x}\left[\frac{f_1(x_1)}{c_1}\right] = -\left[\frac{f_1(x_1)}{c_1}\right]^{\beta}. \tag{34}$$

Note that $f_2(x_2)$ in (33) is what is known as stretched exponential in physics literature. Thus the extended reaction rate integral in (8) is the Mellin convolution of Tsallis nonextensive statistics and stretched exponentials. The starting publication of nonextensive statistical mechanics may be seen from Tsallis (1988). Then proceeding as before, we get,

$$I_{2\beta}^{(\delta)} = \int_0^{\left[\frac{1}{a(1-\beta)}\right]^{\frac{1}{\delta}}} x^{\alpha-1}[1-a(1-\beta)x^{\delta}]^{\frac{1}{1-\beta}} \mathrm{e}^{-bx^{-\rho}} \mathrm{d}x \tag{35}$$

$$= \frac{\Gamma\left(1+\frac{1}{1-\beta}\right)}{\rho\delta[a(1-\beta)]^{\frac{\alpha}{\delta}}} \frac{1}{2\pi i} \int_{c-i\infty}^{c+i\infty} \frac{\Gamma(\frac{s}{\rho})\Gamma(\frac{\alpha}{\delta}+\frac{s}{\delta})[a^{\frac{1}{\delta}}(1-\beta)^{\frac{1}{\delta}} b^{\frac{1}{\rho}}]^{-s}\mathrm{d}s}{\Gamma\left(\frac{\alpha}{\delta}+1+\frac{1}{1-\beta}+\frac{s}{\delta}\right)} \tag{36}$$

$$= \frac{\Gamma\left(1+\frac{1}{1-\beta}\right)}{\rho\delta\left[a(1-\beta)\right]^{\frac{\alpha}{\delta}}} H_{1,2}^{2,0}\left[a^{\frac{1}{\delta}}(1-\beta)^{\frac{1}{\delta}} b^{\frac{1}{\rho}} \middle|_{(0,\frac{1}{\rho}),(\frac{\alpha}{\delta},\frac{1}{\delta})}^{(\frac{\alpha}{\delta}+1+\frac{1}{1-\beta},\frac{1}{\delta})}\right], \tag{37}$$

$a > 0,\ b > 0,\ \alpha > 0, \delta > 0\ \rho > 0,\ \beta < 1,\ \Re(s) > 0,\ \Re(\alpha + s) > 0.$

Make the transformation $\frac{s}{\delta} = s_1$ in (36)

$$I_{2\beta}^{(\delta)} = \frac{\Gamma\left(1+\frac{1}{1-\beta}\right)}{\rho\left[a(1-\beta)\right]^{\frac{\alpha}{\delta}}} \frac{1}{2\pi i} \int_{c-i\infty}^{c+i\infty} \frac{\Gamma(\frac{\delta s_1}{\rho})\Gamma(\frac{\alpha}{\delta}+s_1)[a(1-\beta)b^{\frac{\delta}{\rho}}]^{-s_1}\mathrm{d}s_1}{\Gamma\left(\frac{\alpha}{\delta}+1+\frac{1}{1-\beta}+s_1\right)}. \tag{38}$$

Let us consider the case where $\frac{\delta}{\rho} = m,\ m = 1, 2, \cdots$. Then we get (14).

Lemma 3.1. *As* $\beta \to 1$, $I_{2\beta}^{(\delta)}$ *becomes* $I_2^{(\delta)}$.

Proof.

$$\lim_{\beta\to 1} I_{2\beta}^{(\delta)} = \lim_{\beta\to 1} \frac{1}{2\pi i} \int_{c-i\infty}^{c+i\infty} \frac{\Gamma(\frac{s}{\rho})\Gamma(\frac{\alpha}{\delta}+\frac{s}{\delta})(a^{\frac{1}{\delta}}b^{\frac{1}{\rho}})^{-s}}{\rho\delta a^{\frac{\alpha}{\delta}}} \frac{\Gamma\left(1+\frac{1}{1-\beta}\right)(1-\beta)^{-(\frac{\alpha}{\delta}+\frac{s}{\delta})}}{\Gamma\left(\frac{\alpha}{\delta}+\frac{s}{\delta}+1+\frac{1}{1-\beta}\right)} \mathrm{d}s. \tag{39}$$

Now apply the asymptotic formula for gamma functions, namely, for $|z| \to \infty$ and a is bounded (Mathai 1993),

$$\Gamma(z+a) \to (2\pi)^{\frac{1}{2}} z^{z+a-\frac{1}{2}} \mathrm{e}^{-z}. \tag{40}$$

Apply this to the gamma ratios in (39) by taking z as $\frac{1}{1-\beta}$ and a as 1 and $\frac{\alpha}{\delta}+\frac{s}{\delta}+1$ respectively. Then,

$$\lim_{\beta\to 1} \frac{\Gamma\left(1+\frac{1}{1-\beta}\right)(1-\beta)^{-(\frac{\alpha}{\delta}+\frac{s}{\delta})}}{\Gamma\left(\frac{\alpha}{\delta}+\frac{s}{\delta}+1+\frac{1}{1-\beta}\right)} = 1 \tag{41}$$

Hence

$$\lim_{\beta\to 1} I_{2\beta}^{(\delta)} = I_2^{\delta} \tag{42}$$

which establishes the result.

Case 3.2. $\beta > 1$

$$I_{1\beta}^{(\delta)} = \int_0^\infty x^{\alpha-1}[1+a(\beta-1)x^{\delta}]^{-\frac{1}{\beta-1}} \mathrm{e}^{-bx^{-\rho}} \mathrm{d}x,\ a>0,\ b>0,\ \delta>0,\ \rho>0,\ \beta>1.$$

Here let us take

$$f_1(x_1) = \begin{cases} c_1 x_1^{\alpha}[1+a(\beta-1)x_1{}^{\delta}]^{-\frac{1}{\beta-1}},\ 0<x_1<\infty,\ a>0,\ \delta>0,\ \beta>1 \\ 0, \qquad \text{elsewhere} \end{cases} \tag{43}$$

and

$$f_2(x_2) = \begin{cases} c_2 \mathrm{e}^{-x_2^{\rho}},\ 0<x_2<\infty,\ \rho>0 \\ 0, \qquad \text{elsewhere,} \end{cases} \tag{44}$$

where c_1 and c_2 are the normalizing constants. Observe that $f_1(x_1)$ of (43) is nothing but the superstatistics of Beck and Cohen (2003), and the density in (44) is the stretched exponential. Hence (7) can be looked upon as the Mellin convolution of superstatistics and stretched exponentials. A large number of published articles are there on superstatistics. Then proceeding as before, we get,

$$\begin{aligned} I_{1\beta}^{(\delta)} &= \int_0^\infty x^{\alpha-1}[1+a(\beta-1)x^{\delta}]^{-\frac{1}{\beta-1}} \mathrm{e}^{-bx^{-\rho}} \mathrm{d}x \\ &= K\ H_{1,2}^{2,1}\left[a^{\frac{1}{\delta}}(\beta-1)^{\frac{1}{\delta}} b^{\frac{1}{\rho}} \middle|_{(0,\frac{1}{\rho}),(\frac{\alpha}{\delta},\frac{1}{\delta})}^{(\frac{\alpha}{\delta}+\frac{\beta-2}{\beta-1},\frac{1}{\delta})} \right], \end{aligned} \tag{45}$$

where $K = \frac{1}{\Gamma\left(\frac{1}{\beta-1}\right)\rho\delta[a(\beta-1)]^{\frac{\alpha}{\delta}}}$, $a>0,\ b>0,\ \alpha>0,\ \rho>0,\ \delta>0,\ \beta>1,\ \Re(s)>0,$

$\Re(\alpha+s)>0$. Make the transformation $\frac{s}{\delta}=s_1$ and let $\frac{\delta}{\rho}=m,\ m=1,2,\cdots$ in (45). Then we get (10).

Lemma 3.2. *As* $\beta\to 1$, $I_{1\beta}^{(\delta)}$ *becomes* $I_1^{(\delta)}$.

Particular Case, $\delta=1$

When $\delta=1$, we get the reaction rate integral under the condition $\frac{1}{\rho}=m$, $m=1,2,\cdots$ as,

$$I_{2\beta}=\int_0^{\left[\frac{1}{a(1-\beta)}\right]}x^{\alpha-1}[1-a(1-\beta)x]^{\frac{1}{1-\beta}}\mathrm{e}^{-bx^{-\rho}}\mathrm{d}x,\ a>0,\ b>0,\ \rho>0,\ \beta<1, \tag{46}$$

on evaluating this integral we get (16), and

$$I_{1\beta}=\int_0^{\infty}x^{\alpha-1}[1+a(\beta-1)x]^{\frac{-1}{\beta-1}}\mathrm{e}^{-bx^{-\rho}}\mathrm{d}x,\ a>0,\ b>0,\ \rho>0,\ \beta>1, \tag{47}$$

on evaluating this integral we get (12).

Proof of Theorem 1

For proving the theorem we will make use of the fact that the reaction probability integral as well as the extended reaction probability integrals, as given in (1),(2),(7),(8), under the condition $\delta=m\rho,\ m=1,2,\cdots$ can be written in terms of G-functions. Hence when this condition is satisfied we can invoke the properties of G-functions. It is well known that the G-function defined by

$$G_{p,q}^{m,n}\left[z\Big|_{b_1,,\ldots,b_p}^{a_1,\ldots,a_p}\right]=\frac{1}{2\pi i}\int_L\phi(s)\,z^{-s}\mathrm{d}s, \tag{48}$$

where

$$\phi(s)=\frac{\{\prod_{j=1}^{m}\Gamma(b_j+s)\}\,\{\prod_{j=1}^{n}\Gamma(1-a_j-s)\}}{\{\prod_{j=m+1}^{q}\Gamma(1-b_j-s)\}\,\{\prod_{j=n+1}^{p}\Gamma(a_j+s)\}},$$

$a_j,\ j=1,2,\ldots,p$ and $b_j,\ j=1,2,\ldots,q$ are complex numbers, L is a contour separating the poles of $\Gamma(b_j+s),\ j=1,2,\ldots,m$ from those of $\Gamma(1-a_j-s),\ j=1,2,\ldots,n$, satisfies the following differential equation.

$$\left[(-1)^{p-m-n}z\prod_{j=1}^{p}(\eta-a_j+1)-\prod_{j=1}^{q}(\eta-b_j)\right]G(z)=0,\ \eta=z\frac{\mathrm{d}}{\mathrm{d}z}. \tag{49}$$

This equation is intuitively evident from the following facts:

$$\eta z^{-s} = z\frac{\mathrm{d}}{\mathrm{d}z}(z^{-s}) = (-s)z^{-s};$$

$$(\eta - a_j + 1)z^{-s} = (1 - a_j - s)z^{-s};$$

$$(1 - a_j - s)\Gamma(1 - a_j - s) = \Gamma(2 - a_j - s),$$

(see Mathai 1993). The depleted case of the reaction rate integral

$$I_1^{(\delta)} = \int_0^\infty x^{\alpha-1} e^{-ax^\delta - bx^{-\rho}} \mathrm{d}x, \quad a > 0,\ b > 0,\ \delta > 0,\ \rho > 0 \tag{50}$$

can be expressed in terms of G-functions under the conditions $\frac{\delta}{\rho} = m, m = 1, 2, \cdots$ shown in (4). That is,

$$(2\pi)^{\frac{m-1}{2}} m^{\frac{1}{2}} \rho a^{\frac{\alpha}{m\rho}} I_1^{(m\rho)} = G_{0,m+1}^{m+1,0}\left[\frac{ab^m}{m^m}\bigg|_{0,\frac{1}{m},\cdots,\frac{m-1}{m},\frac{\alpha}{m\rho}}\right], \tag{51}$$

$a > 0,\ b > 0,\ \alpha > 0,\ \delta > 0,\ \rho > 0,\ \frac{\delta}{\rho} = m,\ m = 1, 2, \cdots.$

The G-function in (51), satisfies the differential equation in (3). So the left hand side of (51) also satisfies the differential equation (3). Similar are the proofs of the other theorems and hence deleted.

Acknowledgment The authors would like to thank the Department of Science and Technology, Government of India, New Delhi, for the financial assistance for this work under project-number SR/S4/MS:287/05 and the Centre for Mathematical Sciences for providing all facilities.

References

Beck, C. and Cohen, E.G.D.: Superstatistics. Physica A **322** 267–275 (2003)

Haubold, H.J. and Mathai, A.M.: On nuclear reaction rate theory. Annalen der Physik (Berlin) **41** 380–396 (1984)

Haubold, H.J. and Mathai, A.M.: On thermonuclear reaction rates. Astrophysics and Space Science **258** 185–199 (1998)

Luke, Y.L.: The special functions and their approximations. Academic, New York (1969)

Mathai, A.M., Saxena, R.K.: The H-function with applications in statistics and other disciplines. Wiley Halsted, New York (1978)

Mathai, A.M.: A handbook of generalized special functions for statistical and physical sciences. Oxford University Press, Oxford (1993)

Mathai, A.M. and Haubold, H.J.: Modern problems in nuclear and neutrino astrophysics. Akademie-Verlag, Berlin (1988)

Saxena, R.K.: Some theorems on generalized Laplace transform-1. Proceedings of the national institute of science India **A 26** 400–413 (1960)

Tsallis, C.: Possible generalizations of Boltzmann-Gibbs statistics. Journal of Statistical Physics **52** 479–487 (1988)

Solutions of the Fractional Reaction Equation and the Fractional Diffusion Equation

R.K. Saxena, A.M. Mathai, and H.J. Haubold

Abstract In view of the role of reaction equations in physical problems, the authors derive the explicit solution of a fractional reaction equation of general character, that unifies and extends earlier results. Further, an alternative shorter method based on a result developed by the authors is given to derive the solution of a fractional diffusion equation. Fox functions and Mittag-Leffler functions are used for closed-form representations of the solutions of the respective differential equations.

Keywords Fox function · Mittag-Leffler function · Differential equation · Anomalous reaction & diffusion · Asymptotic expansion · Lévy stable density

1 Introduction

Fractional reaction and diffusion equations involve fractional derivatives with respect to time and space and are studied to describe anomalous reaction and diffusion of dynamical systems with chaotic motion. Fractional reaction equation for Hamiltonian chaos is discussed by Zaslavsky (1994). Solutions and applications of reaction equations are studied by Saichev and Zaslavsky (1997). Solutions of a fractional reaction equation is investigated by Haubold and Mathai (2000) for a simple production-destruction mechanism. This equation was generalized by Saxena et al. (2002). In recent articles, Saxena et al. (2002, 2004a, 2004b) discussed the solution of a number of generalized fractional reaction equations. In the present article, we investigate the solution of a unified fractional reaction equation, which provides unification and extension of results on fractional reaction equations given earlier by Haubold and Mathai (2000) and Saxena et al. (2002, 2004a). We also present the solution of a fractional integral equation discussed by Miller and Ross (1993). Further, an alternative proof of the solution of a fractional diffusion equation given earlier by Kochubei (1990) is investigated, which is based upon a result given by Saxena et al. (2006). Most of the results are obtained in terms of generalized

R.K. Saxena (✉), A.M. Mathai, and H.J. Haubold
Department of Mathematics and Statistics, Jai Narain Vyas University, Jodhpur-342 004, India

H.J. Haubold and A.M. Mathai (eds.), *Proceedings of the Third UN/ESA/NASA Workshop on the International Heliophysical Year 2007 and Basic Space Science*, Astrophysics and Space Science Proceedings, DOI 10.1007/978-3-642-03325-4_7,

Mittag-Leffler functions in elegant and compact form, which are also suitable for numerical computation.

The paper is organized as follows. Section 2 provides the solution of a unified fractional reaction equation while Sect. 3 considers special cases of the equation. A shorter alternative method for the solution of a fractional diffusion equation discussed earlier by Kochubei (1990) is presented in Sect. 4. A series representation and asymptotic expansion of the solution are given in Sect. 5. An H-function representation of a one-sided Lévy stable density is also obtained.

2 Fractional Reaction Equation

In this Section, we present a method based on Laplace transform for deriving the solution of the unified fractional reaction equations.

Theorem 2.1. *If $Re(\nu_j) > 0, a_j > 0, j \in N$, and $f(t)$ be a given function, defined on $\Re_+$, then the equation*

$$N(t) - N_0 f(t) = -\sum_{j=1}^{n} a_j \, {}_0D_t^{-\nu_j} N(t), \tag{1}$$

is solvable and its particular solution is given by

$$N(t) = N_0 \sum_{l=0}^{\infty} (-1)^l \sum_{r_1+\ldots+r_{n-1}=l} \frac{(l)!}{(r_1)!\ldots(r_{n-1})!} \left\{ \prod_{\mu=1}^{n-1} (a_{\mu+1})^{r_\mu} \right\}$$
$$\int_0^t f(u)(t-u)^{\sum_{\mu=1}^{n-1} \nu_{\mu+1} - 1} E^{(l+1)}_{\nu_1, \sum_{\mu=1}^{n-1} \nu_{\mu+1}} [-a_1(t-u)^{\nu_1}] du, \tag{2}$$

where the summation in (2) *is taken over all nonnegative integers $r_1, \ldots, r_n$ such that $r_1 + \ldots + r_{n-1} = l$, and provided that the series and integral in* (2) *are convergent. Here ${}_0D_t^{-\nu_j}$, $j \in N$ are Riemann-Liouville fractional integrals, defined by*

$$ {}_0D_t^{-\nu} f(t) = \frac{1}{\Gamma(\nu)} \int_0^t (t-u)^{\nu-1} f(u) du, \; Re(\nu) > 0, \tag{3}$$

with ${}_0D_t^0 f(t) = f(t)$ (Oldham and Spanier, 1974; Miller and Ross, 1993), $E^{\delta}_{\beta,\gamma}(z)$ is the generalized Mittag-Leffler function, defined by Prabhakar (1971) in terms of series representation as

$$E^{\delta}_{\beta,\gamma}(z) = \sum_{\tau=0}^{\infty} \frac{(\delta)_\tau z^\tau}{\Gamma(\beta\tau+\gamma)(\tau)!} \quad (\beta, \gamma, \delta \in C, Re(\beta) > 0, Re(\gamma) > 0). \tag{4}$$

Proof. By the application of the convolution theorem of the Laplace transform (Erdélyi and Magnus, 1953, p. 259) to (3), we find that

$$L\left\{{}_0D_t^{-\nu}f(t);s\right\} = L\left\{\frac{t^{\nu-1}}{\Gamma(\nu)}\right\}L(f(t)),$$
$$= s^{-\nu}f^{\sim}(s), \tag{5}$$

where $f^{\sim}(s) = \int_0^{\infty} e^{-st}f(t)dt, s \in C, Re(s) > 0$. Applying Laplace transform to (1) and using (5), it gives

$$N^{\sim}(s) = \frac{N_0 f^{\sim}(s)}{1 + a_1 s^{-\nu_1} + \ldots + a_n s^{-\nu_n}} \tag{6}$$
$$= N_0 f^{\sim}(s)\sum_{l=0}^{\infty}(-1)^l \frac{\left(\sum_{j=1}^{n-1} a_{j+1}s^{-\nu_{j+1}}\right)^l}{(1 + a_1 s^{-\nu_1})^{l+1}}, \sum_{j=2}^{n} a_j s^{-\nu_j} < 1 + a_1 s^{-\nu_1}.$$

If we employ the identity (Abramowitz and Stegun, 1968, p.823)

$$(x_1 + \ldots + x_m)^l = \sum_{r_1+\ldots+r_n=l} \frac{(l)!}{(r_1)!\ldots(r_n)!}\prod_{\mu=1}^{m} x_{\mu}^{r_{\mu}}, \tag{7}$$

where the summation is taken over all nonnegative integers, $r_1, \ldots, r_n$, such that $r_1 + \ldots + r_n = l$, then for $|a_1 s^{-\nu_1}| < 1$, (7) transforms into the form

$$N^{\sim}(s) = N_0 f^{\sim}(s)\sum_{l=0}^{\infty}(-1)^l \sum_{\substack{r_1+\ldots+r_{n-1}=l \\ r_1>\ldots r_{n-1}>0}} \frac{(l)!}{(r_1)!\ldots(r_{n-1})!}$$
$$\frac{\left\{\prod_{\mu=1}^{n-1}(a_{\mu+1})^{r_\mu}\right\} s^{-\sum_{\mu=1}^{n-1}\nu_{\mu+1}}}{(1 + a_1 s^{-\nu_1})^{l+1}}. \tag{8}$$

Taking the inverse Laplace transform of (8) by making use of the formula (Kilbas et al. (2004), (12))

$$L^{-1}\left\{s^{-\gamma}(1 - as^{-\beta})^{-\delta};t\right\} = t^{\gamma-1}E_{\beta,\gamma}^{\delta}(at^{\beta}), \tag{9}$$

where $Re(s) > |a|^{1/Re(\gamma)}, Re(\gamma) > 0, Re(s) > 0$, and applying the convolution theorem of the Laplace transform, the result (2) is established.

Remark 2.1. The generalized Mittag-Leffler function defined by (4) is studied by Prabhakar (1971) and Kilbas et al. (2004). Recently this function is used in the theory of finite-size scaling of systems with strong anisotropy and long-range interaction by Chamati and Tonchev (2006).

3 Special Cases

Some special cases of Theorem 2.1 are of interest to be highlighted. If we set $\nu_j = j\nu, a_j = \binom{n}{j}c^{j\nu}(j \in \mathrm{N})$, we obtain

Theorem 3.1. *If $Re(\nu) > 0, c > 0$ and $f(x) \in \Re_+$, then the equation*

$$N(t) - N_0 f(t) = -\sum_{r=1}^{n}\binom{n}{r}c^{\nu r}\, D_t^{-\nu r} N(t), \tag{10}$$

is solvable and its solution has the form

$$N(t) = N_0 \frac{d}{dt}\int_0^t f(u) E_{\nu,1}^{n}[-c^{\nu}(t-u)^{\nu}]du, \tag{11}$$

where $E_{\nu,1}^{n}(x)$ is the generalized Mittag-Leffler function defined by (4) *and provided that the integral* (11) *is convergent.*

When $n = 1$, we obtain the following result given by Hille and Tamarkin (1930).

Corollary 2.1 *Let $Re(\nu) > 0, c > 0$ and let $f(x) \in \Re_+$, then for the solution of the integral equation*

$$N(t) - N_0 f(t) = -c^{\nu}\ {}_0D_t^{-\nu} N(t), \tag{12}$$

holds the following formula

$$N(t) = N_0 \frac{d}{dt}\int_0^t f(u) E_{\nu}[-c^{\nu}(t-u)^{\nu}]du, \tag{13}$$

where $E_{\nu}(z)$ is an entire function of order $\rho = \frac{1}{\nu}$ and type $\sigma = 1$, defined by

$$E_{\nu}(z) = \sum_{\mu=1}^{\infty}\frac{z^{\mu}}{\Gamma(\mu\nu + 1)}, \quad (\nu \in C, Re(\nu) > 0). \tag{14}$$

Note 2.1. The above result has also been given by the authors in a different form (Saxena et al. 2004a, 2004b).

If we set $f(t) = t^{\gamma-1}E_{\nu,\gamma}^{\delta}[-(ct)^{\nu}]$, Theorem 3.1 yields

Corollary 2.2 *Let $Re(\nu) > 0, Re(\gamma) > 0, c > 0$, then for the solution of the reaction equation*

$$N(t) - N_0 t^{\gamma-1}E_{\nu,\gamma}^{\delta}[-(ct)^{\nu}] = -\sum_{r=1}^{n}\binom{n}{r}c^{r\nu}\ {}_0D_t^{-r\nu} n \in N \tag{15}$$

holds the relation

$$N(t) = N_0 t^{\gamma-1} E_{\nu,\gamma}^{\delta+n}[-(ct)^{\nu}], n \in N. \tag{16}$$

For $f(t) = t^{\rho-1}$, *Theorem 3.1 yields the following result*

Corollary 2.3 *If* $Re(\rho) > 0, Re(\nu) > 0, c > 0$, *then for the solution of the equation*

$$N(t) - N_0 t^{\rho-1} = -\sum_{r=1}^{n} \binom{n}{r} c^{r\nu} \ {}_0D_t^{-r\nu} N(t), r \in N, \tag{17}$$

holds the relation

$$N(t) = N_0 t^{\rho-1} \Gamma(\rho) E_{\nu,\rho}^{n}[-(ct)^{\nu}], r \in N. \tag{18}$$

For $n = 1$, (18) *reduces to a result given by Saxena et al. (2002, p. 283,* (15)*). When* $a_j = a^j s^{\nu j}$, *for* $j = 1, \ldots, \nu$, *we obtain*

Theorem 2.2. *Let* $Re(\nu) > 0, a > 0, t > 0, n > 1, |a^{n+1} s^{-(n+1)\nu}| < 1$, *and* $f(x)$ *be a given function defined on* $\Re_+$, *then the equation*

$$N(t) - N_0 f(t) = -\sum_{r=1}^{n} a^r \ {}_0D_t^{-\nu r} N(t), \tag{19}$$

is solvable and its solution is given by

$$\begin{aligned} N(t) = N_0 \Bigg\{ & \frac{d}{dt} \int_0^t f(u) E_{(n+1)\nu,\nu}[a^{n+1}(t-u)^{(n+1)\nu}] du \\ & - a \int_0^t (t-u)^{\nu-1} E_{(n+1)\nu,\nu}[a^{n+1}(t-u)^{(n+1)\nu}] du \Bigg\}, \end{aligned} \tag{20}$$

where $E_{(n+1)\nu,\nu}(z)$ *is the generalized Mittag-Leffler function* $E_{\alpha,\beta}(z)$ *defined as*

$$E_{\alpha,\beta}(z) = \sum_{\mu=1}^{\infty} \frac{z^{\mu}}{\Gamma(\alpha\mu + \beta)}, \ (\alpha, \beta \in C, Re(\alpha) > 0, Re(\beta) > 0) \tag{21}$$

and provided that the integral in (20) *is convergent.*

If we take $\nu^j = j\nu$, *for* $j = 1, \ldots n$, *then it is interesting to note that Theorem 1 yields the following result given by (Miller and Ross, 1993) in a different form:*

Theorem 2.3. *Let* $Re(\nu) > 0, a_j > 0$, *and* $f(x)$ *be a given function defined on* $\Re_+, |a_1 s^{-\nu}| < 1$, *then the fractional reaction equation*

$$N(t) - N_0 f(t) = -\sum_{j=1}^{n} a_j \ {}_0D_t^{-j\nu} N(t), \tag{22}$$

is solvable and has the solution given by

$$N(t) = N_0 \sum_{l=0}^{\infty}(-1)^l \sum_{r_1+\ldots+r_{n-1}=l} \frac{(l)!}{(r_1)!\ldots(r_{n-1})!} \left\{ \prod_{\mu=1}^{n-1} (a_{\mu+1})^{r_\mu} \right\} \tag{23}$$

$$\times \int_0^t f(u)(t-u)^{\sum_{\mu=1}^{n-1} \nu(\mu+1)r_\mu - 1} E^{(l+1)}_{\nu_1, \sum_{\mu=1}^{n-1} \nu(\mu+1)r_\mu}[-a_1(t-u)^{\nu_1}]du,$$

provided that the series and integral in (23) *are convergent.*

4 Fractional Diffusion Equation

In this Section we present an alternative shorter method for deriving the solution of a diffusion equation discussed earlier by Kochubei (1990).

Theorem 4.1. *Consider the Cauchy problem*

$${}_0D_t^{\alpha} N(x,t) = -c^{\nu} \Delta N(x,t), \ (0 < \alpha < 1; x \in \Re^n; \ 0 < t \leq T), \tag{24}$$

with

$$N(x, t = 0) = \delta(x), x \in \Re, \ \lim_{x \to \pm\infty} N(x,t) = 0 \tag{25}$$

${}_0D_t^{\alpha}$ *is the regularized Caputo (1969) partial fractional derivative with respect to t, defined by*

$${}_0D_t^{\alpha} N(x,t) = \frac{1}{\Gamma(1-\alpha)} \left[\frac{\partial}{\partial t} \int_0^t \frac{N(x,s)ds}{(t-x)^{\alpha}} - \frac{N(x,0)}{t^{\alpha}} \right],$$

and Δ *is the Laplacian. The fundamental solution of the above Cauchy problem is given by*

$$N(x,t) = |x|^{-n} \pi^{-\frac{n}{2}} H_{1,2}^{2,0} \left[\frac{|x|^2 t^{-\alpha}}{4c^{\nu}} \Big|_{(n/2,1),(1,1)}^{(1,\alpha)} \right], \tag{26}$$

where $H_{1,2}^{2,0}(.)$ *is the H-function (Mathai and Saxena, 1978).*

Proof. Applying the Laplace transform with respect to t, using the result (Caputo, 1969)

$$L\{{}_0D_t^{\alpha} N(x,t)\} = s^{\alpha} N^{\sim}(x,s) - \sum_{r=0}^{m-1} s^{\alpha-r-1} N^{(r)}(x,0),$$

$$m-1 < \alpha \leq m, \ m \in \mathrm{N},$$

and Fourier transform with respect to x, gives

$$s^{\alpha} N^{\sim^*}(k,s) - s^{\alpha-1} = -c^{\nu}|k|^2 N^{\sim^*}(k,s),$$

where the symbol "$\sim$" indicates the Laplace transform with respect to the time variable t and the symbol "*" the Fourier transform with respect to the space variable x.

Solving for $N^{\sim^*}$, we have

$$N^{\sim^*}(k,s) = \frac{s^{\alpha-1}}{s^{\alpha} + c^{\nu}|k|^2}. \tag{27}$$

By virtue of the following Fourier transform formula (Samko et al. 1990, p. 538, (27.1))

$$\left(F_x\left[|x|^{(2-n)/2}K_{(n-2)/2}(a|x|)\right]\right)(\tau) = \left(\frac{2\pi}{a}\right)^{n/2}\frac{a}{a^2+\tau^2}, (\tau \in \Re^n; n \in \mathrm{N}, a > 0), \tag{28}$$

where the multidimensional Fourier transform with respect to $x \in \Re^n$ is defined by

$$(F_x N)(\tau,t) = \int_{\Re^n} N(x,t)e^{ix\tau}dx \ (\tau \in \Re^n; t > 0) \tag{29}$$

and $K_\nu(.)$ is the modified Bessel function of the second kind, yields

$$\tilde{N}(x,s) = c^{-\nu}s^{\alpha-1}(2\pi)^{-\frac{n}{2}}\left(\frac{|x|c^{\frac{\nu}{2}}}{s^{\alpha}2}\right)^{1-\frac{n}{2}} K_{n-2/2}\left[\frac{|s^{\frac{\alpha}{2}}|x|}{c^{\frac{\nu}{2}}}\right]. \tag{30}$$

In order to invert the Laplace transform, we employ the following result given by the authors Saxena et al. (2006)

$$L^{-1}\{s^{-\rho}K_\nu(zs^{\sigma});t\} = \frac{1}{2}t^{\rho-1}H_{1,2}^{2,0}\left[\frac{z^2t^{-2\sigma}}{4}\middle|_{(\frac{\nu}{2},1)(-\frac{\nu}{2},1)}^{(\rho,2\sigma)}\right], \tag{31}$$

where $K_\nu(x)$ is the modified Bessel function of the second kind, $Re(z^2) > 0$, $Re(s) > 0$. Thus we obtain the solution in the form

$$\begin{aligned} N(x,t) = & \frac{1}{2}(2\pi)^{-\frac{n}{2}}c^{-\frac{\nu}{2}-\frac{n\nu}{4}}|x|^{1-\frac{n}{2}}t^{-\frac{\alpha}{2}-\frac{\alpha n}{4}}s^{\frac{\alpha}{2}+\frac{\alpha n}{4}-1} \\ & H_{1,2}^{2,0}\left[\frac{t^{-\alpha}|x|^2}{4c^{\nu}}\middle|_{(\frac{n-2}{2},1),(\frac{2-n}{4},1)}^{(1-\frac{\alpha}{2}-\frac{\alpha n}{4},\alpha)}\right]. \end{aligned} \tag{32}$$

By virtue of a result in Mathai and Saxena (1978),

$$x^{\sigma}H_{p,q}^{m,n}\left[x\middle|_{(b_q,b_q)}^{(a_p,a_p)}\right] = H_{p,q}^{m,n}\left[x\middle|_{(b_q+\sigma B_q,B_q)}^{(a_p+\sigma A_p,A_p)}\right], \tag{33}$$

the power of the expression $[\{t^{-\nu}|x|^2\}/4c^\nu]$ can be absorbed inside the H-function and consequently we obtain

$$N(x,t) = |\pi^{\frac{1}{2}} x|^{-n}\, H_{1,2}^{2,0}\left[\frac{t^{-\alpha}|x|^2}{4c^\nu}\,\Big|_{(\frac{n}{2},1),(1,1)}^{(1,\alpha)}\right]. \tag{34}$$

Remark 4.1. If we employ the identity (Mathai and Saxena, 1978)

$$H_{p,q}^{m,n}\left[x^\lambda\Big|_{(b_q,B_q)}^{(a_p,A_p)}\right] = \frac{1}{\lambda} H_{p,q}^{m,n}\left[x\Big|_{(b_q,B_q/\lambda)}^{(a_p,A_p/\lambda)}\right], \lambda > 0 \tag{35}$$

the solution given by (32) can be expressed in the form

$$N(x,t) = \frac{1}{\alpha}|\pi^{\frac{1}{2}} x|^{-n}\, H_{1,2}^{2,0}\left[\frac{t^{-1}|x|^{2\alpha}}{(4c^\nu)^{\frac{1}{\alpha}}}\,\Big|_{(\frac{n}{2},\frac{1}{\alpha}),(1,\frac{1}{\alpha})}^{(1,1)}\right], \tag{36}$$

where $\alpha > 0$. We also note that the above form of the solution is due to Schneider and Wyss (1989). There is one importance of our result (32) that it includes the Lévy stable density in terms of the H-function as shown in (34). Similarly, using the identity (33) we arrive at

$$N(x,t) = \frac{1}{2}\pi^{\frac{1}{2}}|x|^{-n} H_{1,2}^{2,0}\left[\frac{t^{-\frac{\alpha}{2}}|x|}{2c^{\frac{\nu}{2}}}\,\Big|_{(\frac{n}{2},\frac{1}{2}),(1,\frac{1}{2})}^{(1,\frac{\alpha}{2})}\right], \tag{37}$$

where n is not an even integer. This form of the H-function is useful in determining its expansion in powers of x. Due to importance of the solution, we also discuss its series representation and behavior.

5 Series Representation of the Solution

Using the series expansion for the H-function given in Mathai and Saxena (1978), it follows that

$$H_{1,2}^{2,0}\left[x\,\Big|_{(\frac{n}{2},\frac{1}{\alpha}),(1,\frac{1}{\alpha})}^{(1,1)}\right] = \frac{1}{2\pi i}\int_L \frac{\Gamma(\frac{n}{2}-\frac{s}{\alpha})\Gamma(1-\frac{s}{\alpha})}{\Gamma(1-s)} x^s\, ds$$
$$= \alpha\left\{\sum_{l=0}^{\infty}\frac{\Gamma(1-\frac{n}{2}-l)(-1)^l x^{\alpha(\frac{n}{2}+l)}}{\Gamma(1-\frac{\alpha n}{2}-\alpha l)(l)!} + \sum_{l=0}^{\infty}\frac{\Gamma(\frac{n}{2}-1-l)(-1)^l x^{\alpha(1+l)}}{\Gamma(1-\alpha-\alpha l)(l!)}\right\}, \tag{38}$$

where n is not an even integer.

Thus for $n = 1$, we find that

$$N(x,t) = \frac{1}{2t^{\frac{\alpha}{2}}} \sum_{l=0}^{\infty} (-1)^l \frac{A^{\frac{l}{2}}}{\Gamma(1-\alpha(l+1)/2)(l!)}, \tag{39}$$

where $A = \frac{x^2}{t^\alpha}$ and the duplication formula for the gamma function is used.

For $n = 2$, the H-function of (37) is singular and in this case, the result is explicitly given by Barkai (2001) in the form

$$N(x,t) \sim \frac{1}{\pi\Gamma(1-\alpha)t^\alpha} ln[\frac{t^{\frac{\alpha}{2}}}{x}]. \tag{40}$$

For $n = 3$, the series expansion is given by

$$N(x,t) = \frac{1}{4\pi t^{3\alpha/2} A^{1/2}} \sum_{l=0}^{\infty} \frac{(-1^l)A^{l/2}}{\Gamma[1-\alpha(1+l/2)]}. \tag{41}$$

From above, it readily follows that for $n = 3$ and $\alpha \neq 1$

$$N(x,t) \sim \frac{1}{x}, \text{ as } x \to \infty. \tag{42}$$

It will not be out of place to mention that the one sided Lévy stable density can be obtained from Laplace inversion formula (31) by virtue of the identity

$$K_{\pm\frac{1}{2}}(x) = \left(\frac{\pi}{2x}\right)^{\frac{1}{2}} e^{-x}, \tag{43}$$

and can be conveniently expressed in terms of the Laplace transform

$$\int_0^\infty e^{-ut}\Phi_\rho(t)dt = e^{-u^\rho}, \quad Re(u) > 0, \; Re(\rho) > 0. \tag{44}$$

The result is

$$\Phi_\rho(t) = \frac{1}{\rho} H_{1,1}^{1,0}\left[\frac{1}{t}\middle|_{(\frac{1}{\rho},\frac{1}{\rho})}^{(1,1)}\right], \; (\rho > 0). \tag{45}$$

Note 5.1. This result is obtained earlier by Schneider and Wyss (1989) by following a different procedure. Asymptotic behavior of $\Phi_\alpha(t)$ is also given by Schneider (1986).

In conclusion, we mention that some of the results derived in this article may find some applications in problems associated with models of long-memory processes driven by Lévy noise and other related problems, see the article by Anh et al. (2002).

Acknowledgment The authors would like to thank the Department of Science and Technology, Government of India, New Delhi, for the financial assistance for this work under project No. SR/S4/MS:287/05 which enabled this collaboration possible.

References

Anh, V.V., Heyde, C.C., Leonenko, N.N.: Dynamic models driven by Lévy noise. Journal of Applied Probability **39** 730–747 (2002)

Abramowitz, M., Stegun, I.A.: Handbook of Mathematical Functions with Formulas, Graphs, and Mathematical Tables. Dover, New York (1968)

Barkai, E.: Fractional Fokker-Planck equation, solution, and application. Physical Review E, **63** 046118 (2001)

Caputo, M.: Elasticita e Dissipazione. Zanichelli, Bologna (1969)

Chamati, H., Tonchev, N.S.: Generalized Mittag-Leffler functions in the theory of finite-size scaling for systems with strong anisotropy and/or long-range interaction. Journal of Physics A. Mathematical and General **39** 469–470 (2006)

Erdélyi, A., Magnus, W., Oberhettinger, F., Tricomi, F.G.: Higher Transcendental Functions. Vol.**1**, McGraw-Hill, New York, Toronto, and London (1953)

Haubold, H.J., Mathai, A.M.: The fractional reaction equation and thermonuclear functions. Astrophysics and Space Science **273** 53–63 (2000)

Hille, E., Tamarkin, J.D.: On the theory of linear integral equations. Annals of Mathematics **31** 479–528 (1930)

Kochubei, A.N.: Diffusion of fractional order. Differential Equations **26** 485–492 (1990)

Kilbas, A.A., Saigo, M., Saxena, R.K.: Generalized Mittag-Leffler function and generalized fractional calculus operators. Integral Transforms and Special Functions **15** 31–49 (2004)

Mathai, A.M., Saxena, R.K.: The H-function with Applications in Statistics and Other Disciplines. Wiley, New York, London, and Sydney (1978)

Miller, K.S., Ross, B.: An Introduction to the Fractional Calculus and Fractional Differential Equations. Wiley, New York (1993)

Oldham, K.B., Spanier, J.: The Fractional Calculus. Theory and Applications of Differentiation and Integration of Arbitrary Order. Academic, New York (1974)

Prabhakar, T.R.: A singular integral equation with a generalized Mittag-Leffler function in the kernel. Yokohama Journal of Mathematics **19** 7–15 (1971)

Saichev, A.I., Zaslavsky, G.M.: Fractional kinetic equations: solutions and applications. Chaos **7** 753–784 (1997)

Samko, S.G., Kilbas, A.A., Marichev, O.I.: Fractional Integrals and Derivatives: Theory and Applications. Gordon and Breach, New York (1990)

Saxena, R.K., Mathai, A.M., Haubold, H.J.: On fractional kinetic equations. Astrophysics and Space Science **282** 281–287 (2002)

Saxena, R.K., Mathai, A.M., Haubold, H.J.: On generalized fractional kinetic equations. Physica A **344** 657–664 (2004a)

Saxena, R.K., Mathai, A.M., Haubold, H.J.: Unified fractional kinetic equation and a fractional diffusion equation. Astrophysics and Space Science **290** 299–310 (2004b)

Saxena, R.K., Mathai, A.M., Haubold, H.J.: Solution of generalized fractional reaction-diffusion equations. Astrophysics and Space Science **305** 305–313 (2006)

Schneider, W.R.: in Stochastic Processes in Classical and Quantum Systems. S. Albeverio, G. Casati, D. Merlini (Eds.), Springer, Berlin (1986)

Schneider, W.R., Wyss, W.: Fractional diffusion and wave equation. Journal of Mathematical Physics **30** 134–144 (1989)

Zaslavsky, G.M.: Fractional kinetic equation for Hamiltonian chaos. Physica D **76** 110–122 (1994)

Astrophysical Applications of Fractional Calculus

Aleksander A. Stanislavsky

Abstract The paradigm of fractional calculus occupies an important place for the macroscopic description of subdiffusion. Its advance in theoretical astrophysics is expected to be very attractive too. In this report we discuss a recent development of the idea to some astrophysical problems. One of them is connected with a random migration of bright points associated with magnetic fields at the solar photosphere. The transport of the bright points has subdiffusive features that require the fractional generalization of the Leighton's model. Another problem is related to the angular distribution of radio beams, being propagated through a medium with random inhomogeneities. The peculiarity of this medium is that radio beams are trapped because of random wave localization. This idea can be useful for the diagnostics of interplanetary and interstellar turbulent media.

Keywords Sun · Magnetic fields · Turbulence · Radio radiation

1 Introduction

At the last two decades, it has become evident that the unfortunate isolation of fractional calculus from exciting developments in physics and astrophysics is beginning to disappear. This is vividly illustrated by the number of recent conferences (for example, Fractional Differentiation and its Applications 2004, Bordeaux; International Symposium on Fractional Calculus, University of Otago, 2006; QEDSP2006; Puschino colloquium "Scattering and scintillation in radioastronomy", 2006; IAU XXVIth General Assembly; UN/ESA/NASA Workshop on Basic Space Science and IHY2007, Tokyo; and so on). The term "fractional calculus" is no new (Oldham and Spanier 1974). Fractional derivatives are almost as old as integer-order definition. Leibniz discussed this subject with L'Hôspital as far back as in 1695. Until recently,

A.A. Stanislavsky (✉)
Institute of Radio Astronomy, 4 Chervonopraporna St., Kharkov, 61002 Ukraine
e-mail: alexstan@ri.kharkov.ua

H.J. Haubold and A.M. Mathai (eds.), *Proceedings of the Third UN/ESA/NASA Workshop on the International Heliophysical Year 2007 and Basic Space Science*, Astrophysics and Space Science Proceedings, DOI 10.1007/978-3-642-03325-4_8,

the area of fractional calculus has primarily been the domain of mathematicians, and it has played a negligible role in physics. Although modern physics is generally described by operators of integer order, the fractional operators can be useful too. Suffice it to recall about the old problems such as the tautochrone problem resolved by Abel (1823) or the Heaviside's analysis (1922) for semi-infinite lossy lines. However, the problems were most likely something exotic. They did not make any recommendations for exploring new physical tasks. The physical sense of fractional operator was foggy enough. The progress in physical applications has been rapid, when the fractional calculus took a relationship with stable distributions in the theory of probability. The connection appeared from the theory of the continuous random walks (CTRW) that occupies an important place for describing many physical phenomena. In such a random walk each subsequent random jump follows after a random waiting time (Montroll and Weiss 1965). In fact, time becomes a sum of positive random intervals that gives a representation about the stochastic arrow of time. If the intervals belong to a stable probabilistic distribution, then their sum is attracted to the same distribution. This is the heart of the generalized Limit Theorem. Formerly the fractional diffusion equations were introduced on a phenomenological base. Nowadays they are derived from the limit processes corresponding to the solutions of the fractional equations. The originality of this approach is that the fractional operators are convolutions (with memory effects), and it follows directly from the features of stable distributions not having a clear connection with any convolution. It is also important that anomalous diffusion is a limit random process in the form of two independent random processes indexed by a common continuous parameter.

The main aim of this paper is to show a profit of the mathematical tools to astrophysical problems. The intention will hold on background of recent achievements of the techniques in physics. In Sect. 2 we briefly describe a circle of physical tasks that have been analyzed starting from fractional calculus and the theory of probability. The problems are connected with anomalous diffusion, superslow relaxation, fractional oscillations, nonlinear phenomena. This approach is quite useful for advancing the results in other sciences, for example, in astrophysics. Some astrophysical applications are considered in Sects. 3 and 4. The future outlook for this study is discussed in Sect. 5.

2 Fractional Calculus in Physics

2.1 *Anomalous Diffusion*

Perhaps no place else in theoretical physics have the techniques of fractional calculus been in the forefront during the last years as in the theory of anomalous diffusion (Metzler and Klafter 2000; Zaslavsky 2002; Stanislavsky 2001, 2002, 2003d; Meerschaert and Scheffler 2004; Piryatinska et al. 2005; Klafter and Sokolov 2005 as well as references therein). Anomalous diffusion plays an important role for many everyday phenomena in physics, astronomy, chemistry, biology, and engineering, where the mean square displacement tends to an increase with time as a power function. Putting it in mathematical terms, let us consider a continuous-time random walk by means of two independent Markov random processes describing the evolution of random variables R_i and T_i, jumps distances and waiting time intervals respectively. If the waiting times T_i belong to an α-stable distribution ($0 < \alpha < 1$), then their sum $n^{-1/\alpha}(T_1 + T_2 + \cdots + T_n)$, $n \in \mathbf{N}$ converges in distribution to a stable law with the same index α (Lamperty 1962). To determine a walker position at the true time t, one needs to find the number of jumps up to time t. This discrete counting process is $\{N_t\}_{t \geq 0} = \max\{n \in \mathbf{N} \mid \sum_{i=1}^{n} T_i \leq t\}$. The continuous limit of $\{N_t\}_{t\geq 0}$ is conventionally denoted by $S(t)$. For a fixed time it represents the first passage of the stochastic time evolution above that time level (Bingham 1971). The random process is nondecreasing, and it can be chosen as a new time clock, the stochastic time arrow (Stanislavsky 2003c). The probability density of the process $S(t)$ has the following Laplace image

$$p^S(t,\tau) = \frac{1}{2\pi j} \int_{Br} e^{ut - \tau u^{\alpha}}\, u^{\alpha-1}\, du = t^{-\alpha} F_{\alpha}(\tau/t^{\alpha}), \tag{1}$$

where Br denotes the Bromwich path. This probability density determines the probability to be at the internal time (or operational time in the book of Feller (1971)) τ on the real time t. The function $F_{\alpha}(z)$, called sometimes Mainardi's one who originally attacked the probability density distribution for anomalous diffusion (Mainardi 1996), can be expanded in a Taylor series. Moreover, it is also an H-function (Fox 1961):

$$F_{\alpha}(z) = H_{11}^{10}\left(z \left| \begin{matrix} (1-\alpha,\alpha) \\ (0,1) \end{matrix} \right.\right) = \sum_{k=0}^{\infty} \frac{(-z)^k}{k!\,\Gamma(1-\alpha(1+k))},$$

where $\Gamma(x)$ is the ordinary gamma function. In the theory of anomalous diffusion the random process $S(t)$ is applied for the subordination of other random processes, for example, Gaussian or Lévy ones (Meerschaert and Scheffler 2004; Stanislavsky 2004c). Since the position vector $\mathbf{r}_t$ is defined by the subordinated process $\{R(S(t))\}_{t\geq 0}$, the probability density of $\mathbf{r}_t$ obeys the integral relationship between the probability densities of the parent process $R(\tau)$ and the directing one $S(t)$:

$$f^{\mathbf{r}_t}(t,x) = \int_0^{\infty} f^R(\tau,x)\, f^S(t,\tau)\, d\tau. \tag{2}$$

Here the probability density $f^R(\tau,x)$ represents the probability to find the parent process $R(\tau)$ at x on the operational time τ. The process $S(t)$ accounts for time, when and how long the walker participates in motion (Baeumer et al. 2005).

The simplest process of anomalous diffusion is obtained when the parent process $\{R(\tau)\}_{\tau\geq 0}$ has the Gaussian probability density $(\pi D\tau)^{-1/2}\exp\{-x^2/(D\tau)\}$, where D is the constant. Then the functional form of the probability density is

$$f^{\mathbf{r}_t}(t,x) = D^{-1/2}\, t^{-\alpha/2}\, F_{\alpha/2}\left(\frac{2\mid x\mid}{\sqrt{D}\, t^{\alpha/2}}\right). \tag{3}$$

As is easily seen (Stanislavsky 2004c), in this case the subordinated process $\{R(S(t))\}$ has finite moments of any order. This property remains valid for the more general case (Stanislavsky 2003b). All the odd moments are equal to zero, and the even moments of $\mathbf{r}_t$ are

$$\langle \mathbf{r}_t^{2n} \rangle = \left(\frac{Dt^\alpha}{4}\right)^n \int_0^\infty z^{2n}\, F_{\alpha/2}(z)\, dz = \frac{(Dt^\alpha)^n\,(2n)!}{4^n\,\Gamma(1+\alpha n)}\,.$$

Thus the process $\mathbf{r}_t$ behaves as subdiffusion ($0 < \alpha < 1$). The limit case with $\alpha = 1$ may be also included in the study, as $T(\tau) = \tau$, and the hitting time process becomes $S(t) = t$. The probability density $f^S(\tau, t)$ reduces to the Dirac δ-function so that after the trivial integration of (2) the function $f^{\mathbf{r}_t}(t, x)$ takes the ordinary Gaussian form. The constant D is interpreted as a generalized diffusion coefficient with dimension $[D] = \text{length}^2/\,\text{time}^\alpha$.

The expression (3) satisfies the fractional Fokker-Planck equation (FPE)

$$f^{\mathbf{r}_t}(t, x) = q(x) + \frac{D}{\Gamma(\alpha)} \int_0^t d\tau (t-\tau)^{\alpha-1} \frac{\partial^2}{\partial x^2} f^{\mathbf{r}_t}(\tau, x), \tag{4}$$

where $q(x)$ is the initial condition. The kernel of this integral equation is a power function resulting in the long-term memory effects in the process $\mathbf{r}_t$. For $\alpha = 1$ (4) becomes the ordinary FPE, where any memory effects are completely absent. To put it in another way, there exists a microscopic time scale which is small in comparison with the observation time t. The microscopic and macroscopic level of the description of such processes are separated in the time scale. Although trajectories have non-differentiable character, their ensemble behavior can be predicted by means of the theoretical prescriptions based on the ordinary differential equations. When the time-scale separation is not valid, memory effects play an important role, i.e. the non-differentiable nature of the microscopic dynamics can be transmitted to the macroscopic level (Stanislavsky 2000). The analysis of anomalous diffusion may be also provided by the Langevin equation (Stanislavsky 2003c; Magdziarz et al. 2007). The competition between normalized and anomalous transport in the presence of subordination is analyzed in (Stanislavsky 2007b).

2.2 *Superslow Relaxation*

Experimental investigations surely have established the relaxation response of amorphous semiconductors and insulators, polymers, molecular solid solutions, glasses to be non-exponential in nature (Jonscher 1996). The nonexponential relaxation results from the asymptotic self-similar properties in the temporal behavior of such systems. As was shown in the paper of Stanislavsky (2004c), the stochastic time arrow can be applied to the general kinetic equation. Denote the transition rates by w defined from microscopic properties of the system (for instance, according to the given Hamiltonian of interaction and the Fermi's golden rule). Then the equation describing a two-state system takes the following form

$$\begin{cases} \tilde{D}^{\alpha} n_{\uparrow}(t) - w\{n_{\downarrow}(t) - n_{\uparrow}(t)\} = 0, \\ \tilde{D}^{\alpha} n_{\downarrow}(t) - w\{n_{\uparrow}(t) - n_{\downarrow}(t)\} = 0, \end{cases} \qquad 0 < \alpha \leq 1, \tag{5}$$

where $n_{\uparrow}$ is the part of components in the state $\uparrow$, $n_{\downarrow}$ the part in the state $\downarrow$, and $\tilde{D}^{\alpha}$ is the α-order fractional derivative with respect to time. Here we use the Caputo derivative (Gorenflo and Mainardi 1997; Podlubny 1999), namely

$$\tilde{D}^{\alpha} x(t) = \frac{1}{\Gamma(n-\alpha)} \int_0^t \frac{x^{(n)}(\tau)}{(t-\tau)^{\alpha+1-n}} \, d\tau, \quad n-1 < \alpha < n,$$

where $x^{(n)}(t) = D^n x(t)$ means the n-derivative of $x(t)$. The relaxation function for the two-state system is written as

$$\phi_{\mathrm{CC}}(t) = 1 - 2n_{\uparrow}(t) = 2n_{\downarrow}(t) - 1 = E_{\alpha}(-2wt^{\alpha}),$$

where $E_{\alpha}(z) = \sum_{n=0}^{\infty} z^n / \Gamma(1 + n\alpha)$ is the one-parameter Mittag-Leffler function. Then the two-state system follows the Cole-Cole law in relaxation. The analysis of this problem can be carried out by another way (Jurlewicz and Weron 2000). It proceeds from the randomization of parameters of distributions that describes the relaxation rates in disordered systems. Its advantages and disadvantages in comparison with the subordination approach are described in (Stanislavsky 2007a).

The evolution of $n_{\uparrow}(t)$ and $n_{\downarrow}$ in (5) can be connected with the Mittag-Leffler distribution. If Z_n denotes the sum of n independent random values with the Mittag-Leffler distribution, then the Laplace transform of $n^{-1/\alpha} Z_n$ is $(1 + s^{\alpha}/n)^{-n}$, which tends to $e^{-s^{\alpha}}$ as n tends to infinity. Following Pillai (1990), this demonstrates an infinite divisibility of the Mittag-Leffler distribution. Due to the power asymptotic form (long tail) the distribution with parameter α is attracted to the stable distribution with exponent α, $0 < \alpha < 1$. The property of the Mittag-Leffler distribution allows one to develop a corresponding stochastic process. The process (called Mittag-Leffler's) arises of subordinating a stable process by a directing gamma process (Pillai 1990). In this case the relaxation function has the Havriliak-Negami (HN) form

$$\phi_{\mathrm{HN}}(t) = 1 - \sum_{k=0}^{\infty} \frac{(-1)^k \Gamma(b+k)}{k! \Gamma(b) \Gamma(1 + ab + ak)} (t/\tau_{\mathrm{HN}})^{ab+ak}, \tag{6}$$

where a, b, τ_{HN} are constant. The one-side Fourier transformation of the relaxation function gives

$$\phi^{\star}_{\mathrm{HN}}(\omega) = \int_0^{\infty} e^{-i\omega t} \left(-\frac{d\phi_{\mathrm{HN}}(t)}{dt} \right) dt = \frac{1}{(1 + (i\omega\tau_{\mathrm{HN}})^a)^b}. \tag{7}$$

This is the well-know HN empirical law. Thus, the HN relaxation can be explained from the subordination approach, if the hitting time process of dipole orientations transforms into the Mittag-Leffler process (Stanislavsky 2003b).

In this connection it should be recalled about the application of the diffusion front properties in stochastic modelling of nonexponential relaxation phenomena. Using the CTRW concept, Gomi and Yonezawa (1995) firstly considered the temporal evolution of diffusion fronts (relaxation properties), when the waiting time distribution is given by a power law type. By Monte-Carlo simulation they came to the Cole-Cole law. The probabilistic background to this idea was done later in the paper of Weron and Kotulski (1996). Later on the consideration of coupled and uncoupled walks gave different forms of relaxation functions (see more details in Stanislavsky 2003a; Magdziarz and Weron 2006). Sometimes the relaxation is represented in time as a product of an exponential and a stretched exponential function (Djordjevič 1986). This can be explained by a two-time scale subordination of random processes (Stanislavsky and Weron 2008).

2.3 *Fractional Oscillations*

The classical Hamiltonian mechanics is formulated in terms of derivatives of integer order. This techniques suggests powerful methods for the analysis of conservative systems in which the total energy does not change. But the real physical world is rather nonconservative because of friction. The derivation of fractional equations of motion from some first physical principles is not an easy matter, as the fractional operator reflects intrinsic dissipative processes. As was shown by Dreisigmeyer and Young (2003), the fractional variational approach does not lead to strictly causal equations of motion because the variational principle utilizes the integration by part, but after this operation the left Riemann-Louville fractional derivative transforms into the right one having a reversal arrow of time. The method of Stanislavsky (2006d) looks at the problem from another point of view. The appearance of the fractional derivative in the equations of motion is conditioned on a peculiar interaction of a physical system with environment. The interaction is taken into account through the temporal variable that represents a sum of random intervals identically distributed. Here the subordination assumes a random waiting time among subsequent deterministic jumps.

Let a Hamiltonian system evolution depend on operational time τ. Then the equation of motion is written as

$$\frac{dq}{d\tau} = \frac{\partial \mathcal{H}}{\partial p}, \qquad \frac{dp}{d\tau} = -\frac{\partial \mathcal{H}}{\partial q}. \tag{8}$$

Consider a fractional dynamical system for which the momentum and the coordinate satisfy the relations

$$p_\alpha(t) = \int_0^\infty p^S(t,\tau)\, p(\tau)\, d\tau,$$
$$q_\alpha(t) = \int_0^\infty p^S(t,\tau)\, q(\tau)\, d\tau.$$

Since the values $\partial\mathcal{H}/\partial p$ and $\partial\mathcal{H}/\partial q$ depend on operational time, we assume that

$$\frac{\partial\mathcal{H}_\alpha}{\partial p_\alpha} = \int_0^\infty p^S(t,\tau)\, \frac{\partial\mathcal{H}}{\partial p}\, d\tau,$$
$$\frac{\partial\mathcal{H}_\alpha}{\partial q_\alpha} = \int_0^\infty p^S(t,\tau)\, \frac{\partial\mathcal{H}}{\partial q}\, d\tau.$$

In this case (8) become fractional:

$$\frac{d^\alpha q_\alpha}{dt^\alpha} = \frac{\partial\mathcal{H}_\alpha}{\partial p_\alpha} = p_\alpha\,, \qquad \frac{d^\alpha p_\alpha}{dt^\alpha} = -\frac{\partial\mathcal{H}_\alpha}{\partial q_\alpha}. \tag{9}$$

When $\alpha = 1$, the generalized equations (9) transform into the ordinary, namely Hamiltonian equations being well known in classical mechanics.

One of the simplest physical models supported by the above-mentioned method is a fractional oscillator (Stanislavsky 2005). Its generalized Hamiltonian takes the form

$$H_\alpha = (p_\alpha^2 + \omega^2 q_\alpha^2)/2, \tag{10}$$

where ω is the circular frequency, q_α and p_α the displacement and the momentum respectively. The value describes the total energy of this dynamical system (Narahari Achar et al. 2001). Although the Hamiltonian (10) is not an explicit function of time, for non-integer values α the system is nonconservative because of the fractional derivative of momentum. Then the Hamiltonian equations for the fractional oscillator are written as

$$\tilde{D}^\alpha q_\alpha = \frac{d^\alpha q_\alpha}{dt^\alpha} = \frac{\partial H_\alpha}{\partial p_\alpha} = p_\alpha\,, \tag{11}$$
$$\tilde{D}^\alpha p_\alpha = \frac{d^\alpha p_\alpha}{dt^\alpha} = -\frac{\partial H_\alpha}{\partial q_\alpha} = -\omega^2 q_\alpha. \tag{12}$$

It follows from this that

$$\tilde{D}^{2\alpha} q_\alpha + \omega^2 q_\alpha = 0 \qquad \text{or} \qquad \tilde{D}^{2\alpha} p_\alpha + \omega^2 p_\alpha = 0. \tag{13}$$

Each of the equations has two independent solutions. Suffice it to solve one of these equations, for example, that determines the coordinate:

$$q_\alpha(t) = A\, E_{2\alpha,1}(-\omega^2 t^{2\alpha}) + B\, \omega t^\alpha\, E_{2\alpha,\,1+\alpha}(-\omega^2 t^{2\alpha}), \tag{14}$$

where A, B are constants, and

$$E_{\mu,\nu}(z) = \sum_{k=0}^{\infty} \frac{z^k}{\Gamma(\mu k + \nu)}, \qquad \mu, \nu > 0,$$

is the two-parameter Mittag-Leffler function (Mathai and Haubold 2007). The momentum of the fractional oscillator is expressed in terms of a fractional time derivative on the coordinate

$$p_\alpha(t) = m\, \tilde{D}^\alpha q_\alpha(t),$$

where m is the generalized mass (Narahari Achar et al. 2001). The approach can be also illustrated with a simple-fractional oscillator under an external force, the coupled fractional oscillators, etc. There exists also a natural extension of this analysis to continuous systems (see details in Stanislavsky 2006d).

2.4 Nonlinear Phenomena

As an example (Zaslavsky et al. 2006), consider the fractional Duffing equation

$$D^\alpha y - y + a y^3 = 0\,, \qquad 1 < \alpha < 2, \tag{15}$$

where $a > 0$ is a constant. The steady states are: $y = 0$, (unstable) and $y = \pm 1/\sqrt{a}$ (stable). Near a stable fixed point by the change $y \to w + 1/\sqrt{a}$ (15) transforms into

$$D^\alpha w + 2w + 3\sqrt{a} w^2 + a w^3 = 0. \tag{16}$$

Close to the stable fixed point we can take a linear equation

$$D^\alpha w_0 + 2 w_0 = 0 \tag{17}$$

with a solution

$$w_0(t) = B E_\alpha(-2t^\alpha), \tag{18}$$

and B is a constant. For $\alpha = 2 - \varepsilon$ and $\varepsilon \ll 1$ expression (18) is well approximated by the relation

$$w_0(t) \approx \frac{2B}{2-\varepsilon} e^{-\sqrt{2}\gamma t} \cos(\sqrt{2}\, t) + O(\varepsilon^2). \tag{19}$$

with $\gamma = -\cos(\pi/\alpha)$. When $\alpha = 2$, (15) and (16) become undamped. The leading term of the frequency of the oscillation w is $\omega_0 = 2^{1/\alpha}$. From the book of Landau and Lifschitz (1976) a nonlinear correction to this frequency is

$$\omega_1 = -\frac{3\, a B^2}{2\sqrt{2}} < 0, \tag{20}$$

when $a B^2 \ll 1$, i.e. $|\omega_1| \ll \omega$.

From (19), we can present w in the form

$$w(t) \approx w_1(t) + \delta w(t) = Be^{-\sqrt{2}\gamma t} \cos\Big[(\omega_0 + \omega_1)t + \beta\Big] + \delta w(t), \tag{21}$$

where β is constant and $\delta w(t)$ is a correction to $\omega_1(t)$.

In fact, the fractional generalization of the considered nonlinear oscillations is reduced to some effective decay of the oscillations similar to the fractional linear oscillator. The rates of the decay can be estimated and, roughly speaking, the larger is the deviation $\varepsilon = 2 - \alpha$, the stronger is the decay. The other fractional nonlinear system was studied in (Stanislavsky 2006b). The model is that the compensation of loss in the linear fractional oscillator by an active device can result in auto-oscillations. Due to the main feature of linear fractional oscillations, namely a finite number of zeros, the limit cycle in such a generator has a short life time depending on the order of fractional derivative. According to Stanislavsky (2006c) the system of three fractional differential equations describes a nonlinear reaction. One of solutions for this system has a peak, and its position in time may be governed by the index choice for the fractional derivative.

2.5 Chaos Versus Long-Term Memory

The treatment of nonlinear dynamics in terms of discrete maps (difference equations produced by numerical methods) is a very important step in studying the qualitative behaviour of continuous systems described by differential equations. When the fractional operator, as a natural generalization of the ordinary differentiation and integration, depends on time, it is characterized by long-term memory effects corresponding to intrinsic dissipative processes in the physical systems. In the application to discrete maps this means that their present state evolution depends on all past states with a power weight.

The mapping $x_{n+1} = f(x_n)$ does not have any memory. Using the weights from the robust algorithm of numerical fractional integration (Diethelm et al. 2005) such as

$$c_i^{(n)} = \begin{cases} (1+\alpha)n^\alpha - n^{\alpha+1} + (n-1)^{\alpha+1}\,, & \text{if } i = 0; \\ 1\,, & \text{if } i = n; \\ (n-i+1)^{\alpha+1} - 2(n-i)^{\alpha+1} + & \\ +(n-i-1)^{\alpha+1}\,, & \text{if } 0 < i < n-1, \end{cases}$$

the normalized mapping with long-term memory effects is written as

$$x_{n+1} = \frac{1}{(1+\alpha)n^\alpha} \sum_{i=0}^{n} c_i^{(n)} f(x_i), \tag{22}$$

where $f(x)$ is an arbitrary function suitable for the discrete map. The memory effects are governed by means of the parameter α. If $\alpha = 0$, the memory is absent as for ordinary logistic maps. If $\alpha = 1$, the map has the full memory, i.e. each new state of the discrete system sticks to the system's memory and has the same action upon next states as all the others in memory. When $0 < \alpha < 1$, the memory effects have a long-term dependence. The numerical simulations (Stanislavsky 2006a) show that the chaotic sequences which are presented in the discrete systems without any memory are squeezed out. As the memory mounts, chaos and its traces disappear. The magnitude $\alpha_{\text{crit}} \approx 0.155$ looks like a "critical point" that indicates a transition from chaos to order. It should be noticed that the fine structure in the iterates for the maps is just washed out because of the memory effects. Nevertheless, for sufficiently small values of α there is still a remarkable transition to chaos.

The similar situation takes the place in continuous systems with long-term memory. In the paper (Zaslavsky et al. 2006) it is shown that the fractional nonlinear oscillator behaves like the stochastic attractor in phase space, being periodically perturbed. The memory effect dissipation leads to a degradation of the attractor structure. If the equations with fractional derivatives slightly different from the integer ones by $\varepsilon = 2 - \alpha \ll 1$, then one can be fairly easy interpreted through the regular equations with a dissipation.

In the next two sections we will consider some applications of the subordination analysis in astrophysical problems. This is something like a concrete test of strength.

3 Subdiffusive Transport of Magnetic Points on the Sun

Due to the complicated character of the solar convection, the heliospheric magnetic field has a complex behaviour like random. Leighton (1964) suggested to study the migration of magnetic regions on the Sun as a random walk on the solar surface. However, the modern precise observations of this migration have shown (Hagenaar et al. 1999) that the transport of magnetic regions in the intergranular lanes with times less than 20 min has a non-Gaussian (subdiffusive) character. The point is that the migration of the magnetic flux from a point to a point tends to accumulate at sinks of the flow field (Simon et al. 1995). The sinks displace randomly. Therefore, magnetic elements stick before the next jump (Cadavid et al. 1999). The Leighton's model clearly leads to a diffusion equation. How to describe the random walk with long-term memory on the solar surface?

Apply the approach, developed in Sect. 2, to this problem.The distribution $g_\alpha(\theta, \phi, t)$ of magnetic field footpoints on the solar spherical surface of radius $R_\odot$ obeys the two-dimensional equation in spherical coordinates

$$\frac{\partial^\alpha g_\alpha}{\partial t^\alpha} - \frac{g_\alpha(\theta, \phi, 0)\, t^{-\alpha}}{\Gamma(1-\alpha)} = \frac{1}{R_\odot \sin\theta}\frac{\partial}{\partial\theta}\left(\frac{\kappa \sin\theta}{R_\odot}\frac{\partial g_\alpha}{\partial\theta}\right) + \frac{1}{R_\odot \sin\theta}\frac{\partial}{\partial\phi}\left(\frac{\kappa}{R_\odot \sin\theta}\frac{\partial g}{\partial\phi}\right), \tag{23}$$

where κ is the subdiffusion coefficient with the following dimension $[\kappa] = 1/\text{time}^{\alpha}$, and $\partial^{\alpha}/\partial t^{\alpha}$ denotes the Liouville-Riemann fractional differential operator of order α (Samko et al. 1993). The Liouville-Riemann fractional differential operator of integer order is nothing else but the ordinary derivative. The solution of (23) can be obtained by a separation ansatz in terms of the spherical harmonics $Y_{mn}(\theta, \phi)$, namely

$$g_{\alpha}(\theta, \phi, t) = \sum_{n=0}^{\infty} \sum_{m=-n}^{m=n} a_{mn}\, Y_{mn}(\theta, \phi)\, E_{\alpha}\left(-\frac{\kappa}{R_{\odot}^{2}}\, n(n+1)\, t^{\alpha}\right). \tag{24}$$

If one takes an impulse release of footpoints at the location (θ_0, ϕ_0), the initial condition is expressed in terms of the Dirac δ-function:

$$g_{\alpha}(\theta, \phi, 0) = \frac{1}{\sin\theta}\, \delta(\phi - \phi_0)\, \delta(\theta - \theta_0).$$

Using the completeness relationship of spherical harmonics states, the amplitude coefficients a_{mn} are the complex conjugate of the spherical harmonics $Y_{mn}^{*}(\theta_0, \phi_0)$. For $t \to \infty$ the distribution approaches asymptotically the limit

$$\lim_{t\to\infty} g_{\alpha}(\theta, \phi, t) = \frac{1}{4\pi}.$$

This result may have expected all along, as it is nothing else but the homogeneous distribution on the sphere in the equilibrium. After integrating the expression (24) from $\phi = 0$ to 2π, the solution of (23) only will depend on the polar angle θ and time t:

$$g_{\alpha}(\theta, t) = \sum_{n=0}^{\infty} \frac{2n+1}{2}\, P_n(\cos\theta_0)\, P_n(\cos\theta)\, E_{\alpha}\left(-\frac{\kappa}{R_{\odot}^{2}}\, n(n+1)\, t^{\alpha}\right), \tag{25}$$

where $P_n(y)$ is the Legendre polynomial of degree n. All the odd moments of the distribution are equal to zero, but the even moments are not. In particular, the first even moment takes the form

$$\overline{\sin^2\theta} = \frac{2}{3}\left(1 - E_{\alpha}\left(-\frac{6\kappa}{R_{\odot}^{2}}\, t^{\alpha}\right)\right).$$

Since the subdiffusion motion of magnetic footpoints is characterized by a small value $\kappa/R_{\odot}^{2}$ (about 0.001 in order of magnitude), we arrive at the relation

$$\overline{\theta^2} \approx \frac{4\kappa}{\Gamma(1+\alpha)\, R_{\odot}^{2}}\, t^{\alpha}. \tag{26}$$

It is obvious that for $\alpha = 1$ this description leads to the Leighton's model. The comparison of the subdiffusive model with experimental data may be carried out in the same manner as this is the case in the work of Cadavid et al. (1999). Thus, the Leighton's model may be generalized. Although the experimental data for the motion of bright points (associated with magnetic fields at the photosphere) show a two-time scaling, the constructed model is applicable for the analysis of diffusive processes that change their character, from subdiffusion to normal one on various temporal scales.

4 Beam Propagation in Strong Turbulent Media

The wave nature of diffusing particles can lead to a complete halt of diffusion because of interference of waves propagating in reciprocal multiple scattering paths, and therefore, the particles stay close to their initial place. The phenomenon is called the Anderson localization (Anderson 1958) who first predicted it for explaining the metal-insulator transition in electronic systems. The localization concept may be also applied to classical wave systems (Sheng 1995). One of the interesting applications is a beam propagation in strong turbulent media, where there may exist the effect caused by wave localization at random locations and random times. This leads to traps of beams by inhomogeneities in such a medium, and the propagation of beams contains random jerking. Consider the problem in more details.

If a beam propagates through a random medium, then it deflects at random directions. The angle θ of deviation of a beam from its initial direction is characterized by a probability density $W_\alpha(\theta,\sigma)$, where σ is the path traveled by the beam. Let us derive an integro-differential equation for the probability density. In contrast to rotational Brownian motion, the random walks analyzed here consist of random angle jumps $\Delta\theta_i$ at points separated by segments of random length $\Delta\sigma_i$. Assume that the angle jumps are independent random variables belonging the domain of attraction of Gaussian probability distribution, and the random segment lengths $\Delta\sigma_i$ are also identically distributed independent random variables, with stable distribution by an index α. Since $\Delta\sigma_i$ is a nonnegative quantity, this distribution is totally asymmetric, and $0 < \alpha \leq 1$. On account of convergence of distributions we can definitely pass from the discrete model to a continuous limit. According to Stanislavsky (2004b) this leads to the diffusion equation

$$W_\alpha(\theta,\sigma) - W_\alpha(\theta,0) = \int_0^\sigma \frac{D}{\Gamma(\alpha)\sin\theta}\frac{\partial}{\partial\theta}\left(\sin\theta\frac{\partial W_\alpha(\theta,\sigma')}{\partial\theta}\right)(\sigma-\sigma')^{\alpha-1}\,d\sigma'\,,$$

where D is a diffusion coefficient. The solution to this equation can be expressed as an integral transform of the probability density associated with rotational Brownian motion:

$$W_\alpha(\theta,\sigma) = \int_0^\infty F_\alpha(z)\,W_1(\theta,\sigma^\alpha z)\,dz\,.$$

The diffusion equation yields the mean

$$\overline{\cos\theta} = E_\alpha(-2D\sigma^\alpha).$$

Following the method developed by Chernov (1953), one can find the mean square of the distance r from the starting point to the observation point reached by the beam that has travelled an intricate path of length σ through the medium:

$$\overline{r^2} = \frac{\sigma^\alpha}{D\Gamma(\alpha+1)} - \frac{1}{2D^2}\Big(1 - E_\alpha(-2D\sigma^\alpha)\Big). \tag{27}$$

If $D\sigma \ll 1$, then

$$\overline{r^2} \approx 2\sigma^{2\alpha}\left(\frac{1}{\Gamma(2\alpha+1)} - \frac{2D\sigma^\alpha}{\Gamma(3\alpha+1)}\right). \tag{28}$$

If the z axis of a polar coordinate system is aligned with the initial beam direction, then the mean square of the distance passed by the beam along this axis is given by the formula

$$\overline{z^2} = \frac{1}{3D}\left[\frac{\sigma^\alpha}{\Gamma(\alpha+1)} - \frac{1}{6D}\Big(1 - E_\alpha(-6D\sigma^\alpha)\Big)\right]. \tag{29}$$

If $D\sigma$ is enough small, then

$$\overline{z^2} \approx 2\sigma^{2\alpha}\left[\frac{1}{\Gamma(2\alpha+1)} - \frac{6D\sigma^\alpha}{\Gamma(3\alpha+1)}\right]. \tag{30}$$

Now, the mean square deviation of the beam from its initial direction can be calculated by combining (27) with (29):

$$\overline{\rho^2} = \overline{r^2} - \overline{z^2} = \frac{2\sigma^\alpha}{3D\Gamma(\alpha+1)} - \frac{1}{2D^2}\Big(1 - E_\alpha(-2D\sigma^\alpha)\Big) + \frac{1}{18D^2}\Big(1 - E_\alpha(-6D\sigma^\alpha)\Big). \tag{31}$$

Then the ordinary "3/2 law" (Chernov 1953) takes a generalized form

$$\sqrt{\overline{\rho^2}} \approx \frac{2\sqrt{2}}{\sqrt{\Gamma(3\alpha+1)}} D^{1/2}\sigma^{3\alpha/2}. \tag{32}$$

The mean squares given by (27), (29), and (31) increase as σ^α at large σ. The case of $\alpha = 1$ corresponds to normal diffusion without wave localization. Thus, the approach developed here embraces classical results of the theory of beam propagation in a randomly inhomogeneous medium (Chernov 1960). As has been shown by Stanislavsky (2007c), the character of subordination can modify the mean square deviation of the beam from its initial direction noticeably, and the subordinators may have various forms. This feature is useful for the diagnostics of interplanetary and interstellar turbulent media.

5 Conclusions

Normal diffusion as a model is often used in astrophysics, but it does not always justify hopes. For example, the classical theory of scintillations assumes that the individual angular increments are independent and identically Gaussian distributed random variables. However, the pulse profiles predicted by the theory disagree with the observable results for the distant pulsars (Sutton 1971). Recently Boldyrev and Gwinn (2003, 2005) have shown that the time broadening of the radio pulses from the distant pulsars can arise from non-Gaussian statistics of interstellar electron-density. The problem requires the Lévy-flights scenario. The Lévy statistics contains larger-than-rare events, and its variance diverges. The Lévy flights in time has a direct relationship to subdiffusion. It originates from a subordination of one random process by another, and the latter is governed by a Lévy process. The subdiffusion stems from weak localization effects. In this paper we have shown capabilities of this approach in physics and have suggested some possible applications in astrophysics. This presentation serves but to convey a deeper impression of the future progress expected in astrophysics due to fractional calculus.

Acknowledgements The author thanks Professor Hans Haubold for his help and support of this work as well as the National Observatory of Japan for hospitality during the UN/ESA/NASA Workshop on Basic Space Science and the International Heliophysical Year 2007. This research was partly done with the project INTAS-03-5727.

References

Anderson, P.W.: Absence of Diffusion in Certain Random Lattices. Phys. Rev. 109, 1492–1505 (1958)

Baeumer, B., Benson, D.A., Meerschaert, M.M.: Advectionand dispersion in time and space. Physica A 350, 245–262 (2005)

Bingham, N.: Limit theorems for occupation times of Markov processes. Z. Wharsch. verw. Geb. 17, 1–22 (1971)

Boldyrev, S., Gwinn, C.R.: Scintillations and Lévy flights through the interstellar medium. Astrophys. J. 584, 791–796 (2003); Ibid.: Radio-wave propagation in the non-Gaussian interstellar medium. Astrophys. J. 624, 213–231 (2005)

Cadavid, A.C., Lawrence, J.K., Ruzmaikin, A.A.: Anomalous diffusion of solar magnetic elements. Astrophys. J. 521, 844–850 (1999)

Chernov, L.A.: Sound Propagation in a Randomly Inhomogeneous Medium. Zh. Eksp. Teor. Fiz. 24, 210–213 (1953)

Chernov, L.A.: Wave Propagation in a Random Medium. McGraw-Hill, New York (1960)

Diethelm, K., Ford, N.J., Freed A.D., Luchko Yu.: Algorithms for the fractional calculus: a selection of numerical methods. Comput. Methods Appl. Mech. Eng. 194, 743–773 (2005)
Djordjevič, Z.B.: Scaling observation in the trapping reaction. In: Pietronero, L., Tossati, E. (eds.) Fractals in Physics. Proceeding of Sixth Trieste International Symposium, pp. 581–585. North-Holland, Amsterdam (1986)
Dreisigmeyer, D.W., Young, P.M.: Nonconservative Lagrangian mechanics: a generalized function approach. J. Phys. A: Math. Gen. 36, 8297–8310 (2003)
Feller, W.: An Introduction to Probability Theory and Its Aplications. Wiley, New York (1971)
Fox, C.: The G and H-functions as symmetrical Fourier kernels. Trans. Am. Math. Soc. 98, 395–429 (1961)
Gomi S., Yonezawa, F.: Anomalous Relaxation in the Fractal Time Random Walk Model. Phys. Rev. Lett. 74, 4125–4128 (1995)
Gorenflo, R., Mainardi, F.: Fractional calculus: integral and differential equations of fractional order. In: Carpinteri, A., Mainardi, F. (eds.) Fractals and Fractional Calculus in Continuum Mechanics, pp. 223–276. Springer, New York (1997)
Hagenaar, H.J., Schrijver, C.J., Title, A.M., Shine, R.A.: Dispersal of magnetic flux in the quiet solar photosphere. Astrophys. J. 511, 932–944 (1999)
Heaviside, O.: Electromagnetic Theory. Vol. II, Benn, London (1922), 547 p. (Reprinted New York: Chelsea Publishing 1971)
Jonscher, A.K.: Universal Relaxation Law. Chelsea Dielectric Press, London (1996)
Jurlewicz, A., Weron, K.: Infinitely Divisible Waiting-Time Distributions Undelying the Empirical Relaxation Responses. Acta Physiol. Pol. B 31, 1077–1084 (2000)
Klafter, J., Sokolov, I.M.: Anomalous Diffusion spreads its wings. Physics World, August 2005, pp. 29–32
Lamperty, J.: Semi-stable stochastic processes. Trans. Am. Math. Soc. 104, 62–78 (1962)
Landau, L.D., Lifschitz E.M.: Mechanics. 3rd ed., Pergamon Press, Oxford, England, (1976)
Leighton, R.B.: Transport of magnetic field on the Sun. Astrophys. J. 140, 1547–1562 (1964)
Magdziarz, M., Weron, K.: Anomalous diffusion schemes underlying the stretched exponential relaxation. The role of subordinators. Acta Physiol. Pol. B 37(5), 1617–1625 (2006)
Magdziarz, M., Weron A., Weron, K.: Fractional Fokker-Planck dynamics: Stochastic representation and computer simulations. Phys. Rev. E 75(2), 016708 (2007)
Mainardi, F.: Fractional Relaxation-Oscillation and Fractional Diffusion-Wave Phenomena. Chaos Solitons Fractals 7(9), 1461–1477 (1996)
Mathai, A.M., Haubold, H.J.: Special Functions for applied scientists. Publication No. 35, Center for Mathematical Sciences Pala Campus, Arunapuram P.O., Palai, Kerala - 686574, India, February 2007, 416 p
Meerschaert, M.M., Scheffler, H.-P.: Limit theorems for continuous-time random walks with infinite mean waiting time, J. Appl. Probab. 41, 623–638 (2004)
Metzler, R., Klafter, J.: The random walk's guide to anomalous diffusion: A fractional dynamics approach. Phys. Rep. 339, 1–77 (2000)
Montroll, E.W., Weiss, G.H.: Random Walks om Latices. II. J. Math. Phys. 6(2), 167–181 (1965)
Narahari Achar, B.N., Hanneken, J.W., Enck, T., Clarke, T.: Dynamics of the fractional oscillator. Physica A 297, 361–367 (2001)
Oldham, K.B., Spanier, J.: The Fractional Calculus. Academic Press, San Diego (1974)
Pillai R.N: On Mittag-Leffler functions and related distributions. Ann. Inst. Statist. Math. 42(1), 157–161 (1990)
Piryatinska A., Saichev, A.I., Woyczynski, W.A.: Models of anomalous diffusion: the subdiffusive case. Physica A 349, 375–420 (2005)
Podlubny, I.: Fractional Differential Equations. Academic Press, San Diego (1999)
Samko, S.G., Kilbas, A.A., Marichev, O.I.: Fractional Integrals and Derivatives – Theory and Applications. Gordon and Breach, New York (1993)
Sheng, P.: Introduction to Wave Scattering, Localization and Mesoscopic Phenomena. Academic Press, New York (1995)

Simon, G.W., Title, A.M., Weiss, N.O.: Kinematic model of supergranular diffusion on the Sun. Astrophys. J. 442, 886–897 (1995)

Stanislavsky, A.A.: On the influence of fluctuations of the antenna effective area on radio astronomy observations in decameter wavelengths. In: Abstract Book of XXVI General Assembly of URSI, Toronto, Canada, 14–21 August 1999, p. 604

Stanislavsky, A.A.: Memory effects and macroscopic manifestation of randomness. Phys. Rev. E 61, 4752–4759 (2000)

Stanislavsky, A.A.: Antenna effective area fluctuations and $1/f^{\alpha}$-noise. Chaos Solitons Fractals 12(13), 2407–2410 (2001)

Stanislavsky, A.A., Weron, K.: Exact solution of averaging procedure over the Cantor set. Physica A 303(1–2), 57–66 (2002)

Stanislavsky, A.A.: Subordinated random walk approach to anomalous relaxation in disordered systems. Acta Physiol. Pol. B, 3649–3660 (2003a)

Stanislavsky, A.A.: Subordinated Brownian motion and its fractional Fokker-Planck equation. Physica Scr. 67(4), 265–268 (2003b)

Stanislavsky, A.A.: Fractional dynamics from the ordinary Langevin equation. Phys. Rev. E 67, 021111 (2003c)

Stanislavsky, A.A.: Black-Scholes model under subordination. Physica A 318(3–4), 469–474 (2003d)

Stanislavsky, A.A.: Fractional oscillator. Phys. Rev. E 70, 051103 (2004a)

Stanislavsky, A.A.: Beam propagation in a randomly inhomogeneous medium. J. Exp. Theor. Phys. 98(4), 705–706 (2004b)

Stanislavsky, A.A.: Probability interpretation of the integral of fractional order. Theor. Math. Phys. 138(3), 418–431 (2004c)

Stanislavsky, A.A.: Twist of fractional oscillations. Physica A 354, 101–110 (2005)

Stanislavsky, A.A.: Long-term memory contribution as applied to the motion of discrete dynamical systems. Chaos 16(4), 043105 (2006a)

Stanislavsky, A.A.: The peculiarity of self-excited oscillations in fractional systems. Acta Physiol. Pol. B 37(2), 319–329 (2006b)

Stanislavsky, A.A.: Nonlinear reaction with fractional dynamics. Appl Math Comput 174(2), 1122–1134 (2006c)

Stanislavsky, A.A.: Hamiltonian formalism of fractional systems. Europ. Phys. J. B 49(1), 93–101 (2006d)

Stanislavsky, A.A., Mukha, D.V.: Transport of solar magnetic elements in the intergranular lanes. In: Abstract Book of IAU XXVIth General Assembly, Prague, 14–25 August 2006, p. 284

Stanislavsky, A.A.: Subdiffusion of beams through interplanetary and interstellar media,. In: Abstracts of colloquium "Scattering and scintillation in radioastronomy", 19–23 June 2006, Pushchino, Moscow region, Russia, p. 19

Stanislavsky, A.A.: The Stochastic Nature of Complexity Evolution in the Fractional Systems. Chaos Solitons Fractals 34(1), 51–61 (2007a)

Stanislavsky, A.A.: Effect of temporal randomization to the interaction of normalized and anomalous transport. Probl. At. Sci Technol, 3, ch. II, 340–342 (2007b)

Stanislavsky, A.A.: Subdiffusion of beams through interplanetary and interstellar media. Astron. Astrophys. Transactions 26(6), 655–658 (2007c)

Stanislavsky, A.A., Weron K.: Two-time scale subordination in physical processes with long-term memory. Ann. Phys. 323(3), 643–653 (2008)

Sutton, J.M.: Scattering of pulsar radiation in the interstellar medium. Mon. Not. R. Astron. Soc. 155, 51–64 (1971)

Weron K., Kotulski M.: On the Cole-Cole relaxation function and related Mittag-Leffler distribution. Physica A 232, 180–188 (1996)

Zaslavsky, G.M.: Chaos, fractional kinetics, and anomalous transport. Phys. Rep. 371, 461–580 (2002)

Zaslavsky, G.M., Stanislavsky, A.A., Edelman, M.: Chaotic and pseudochaotic attractors of perturbed fractional oscillator. Chaos 16(1), 013102 (2006)

Generalized Mittag-Leffler Distributions and Processes for Applications in Astrophysics and Time Series Modeling

Kanichukattu Korakutty Jose, Padmini Uma, Vanaja Seetha Lekshmi, and Hans Joachim Haubold

Abstract Geometric generalized Mittag-Leffler distributions having the Laplace transform $\frac{1}{1+\beta \log(1+t^{\alpha})}, 0 < \alpha \leq 2, \beta > 0$ is introduced and its properties are discussed. Autoregressive processes with Mittag-Leffler and geometric generalized Mittag-Leffler marginal distributions are developed. Haubold and Mathai (Astrophysics and Space Science 273 53–63, 2000) derived a closed form representation of the fractional kinetic equation and thermonuclear function in terms of Mittag-Leffler function. Saxena et al. (2002; Astrophysics and Space Science 209 299–310 2004a; Physica A 344 657–664 2004b) extended the result and derived the solutions of a number of fractional kinetic equations in terms of generalized Mittag-Leffler functions. These results are useful in explaining various fundamental laws of physics. Here we develop first-order autoregressive time series models and the properties are explored. The results have applications in various areas like astrophysics, space sciences, meteorology, financial modeling and reliability modeling.

Keywords Autoregressive process · α-Laplace distribution · Geometric infinite divisibility · Geometric generalized Mittag-Leffler distribution · Generalized Mittag-Leffler distribution · Self decomposability · Time series modeling · Financial modeling

1 Introduction

Recently Mittag-Leffler functions and distributions have received the attention of mathematicians, statisticians and scientists in physical and chemical sciences. Pillai (1990) introduced the Mittag-Leffler distribution in terms of Mittag-Leffler functions. Jayakumar and Pillai (1993) developed a first order autoregressive process with Mittag-Leffler marginal distribution. Fujita (1993) discussed a generalization of the results of Pillai. Jose and Seetha Lekshmi (1997) developed geometric

K.K. Jose (✉), P. Uma, V.S. Lekshmi, and H.J. Haubold
Department of Statistics, St. Thomas College, Pala, M.G. University, Kerala-686 574, India
e-mail: kkjstc@gmail.com;kkjstc@rediffmail.com

H.J. Haubold and A.M. Mathai (eds.), *Proceedings of the Third UN/ESA/NASA Workshop on the International Heliophysical Year 2007 and Basic Space Science*, Astrophysics and Space Science Proceedings, DOI 10.1007/978-3-642-03325-4_9,

exponential distribution and Seetha Lekshmi and Jose (2002, 2003) extended the results to obtain geometric Mittag-Leffler distributions. Jayakumar and Ajitha (2003) obtained various results on geometric Mittag-Leffler distributions. Seetha Lekshmi and Jose (2004) extend the concept to obtain geometric α-Laplace processes.

A discrete version of the Mittag-Leffler distribution was introduced by Pillai and Jayakumar (1995). Lin (1998a,b, 2001) obtained various characterizations of the Mittag-Leffler distributions and refers to them as positive Linnik laws. In physics, Haubold and Mathai (2000) derived a closed form representation of the fractional kinetic equation and thermonuclear function in terms of Mittag-Leffler function. Saxena et al. (2004a,b) extended the result and derived the solutions of a number of fractional kinetic equations in terms of generalized Mittag-Leffler functions and obtained the solution of a unified form of generalized fractional kinetic equations, which provides the unification and extension of the earlier results. Such behaviors occur frequently in chemistry, thermodynamical and statistical analysis. In all such situations the solutions can be expressed in terms of generalized Mittag-Leffler functions. Weron and Kotulski (1996) use Mittag-Leffler distribution in explaining Cole–Cole relaxation.

The function $E_\alpha(z) = \sum_{k=0}^{\infty}\left[\frac{z^k}{\Gamma(1+ak)}\right]$ was first introduced by Mittag-Leffler in 1903 (Erdélyi 1955). Many properties of the function follow from Mittag-Leffler integral representation

$$E_\alpha(z) = \frac{1}{2\pi i}\int_C \frac{t^{\alpha-1}e^t}{t^\alpha - z}dt.$$

where the path of integration C is a loop which starts and ends at $-\infty$ and encircles the circular disc $|t| \le z^{\frac{1}{\alpha}}$. Pillai (1990) proved that $F_\alpha(x) = 1 - E_\alpha(-x^\alpha), 0 < \alpha \le 1$ are distribution functions, having the Laplace transform $\psi(t) = (1+t^\alpha)^{-1}, t \ge 0$ which is completely monotone for $0 < \alpha \le 1$. He called $F_\alpha(x)$, for $0 < \alpha \le 1$, a Mittag-Leffler distribution. The Mittag-Leffler distribution is a generalization of the exponential distribution, since for $\alpha = 1$, we get exponential distribution. Pillai (1990) has shown that $F_\alpha(x)$ is geometrically infinitely divisible (g.i.d.) and is in the domain of attraction of stable laws.

Pillai (1985) developed α-Laplace distribution with characteristic function given by $(1 + |t|^\alpha)^{-1}$; $0 < \alpha \le 2$. This distribution is also known as Linnik distribution. Using a simple characterization of this distribution given by Linnik (1962), Anderson and Arnold (1993) constructed a discrete-time process having a stationary Linnik distribution. These Linnik models appear to be viable alternatives to stable processes as models for temporal changes in stock prices. Pakes (1998) gave a mixture representation for symmetric generalized Linnik laws as $A_{\alpha,\nu} \cong (\gamma(\nu))^{\frac{1}{\alpha}} S_\alpha$, where $A_{\alpha,\nu}$ denotes a random variable having the characteristic function $(1+|t|^\alpha)^{-\nu}, \nu > 0$. This defines the generalized symmetric Linnik law $SLi(\alpha,\nu)$. Kotz and Ostrovskii (1996) proved the following mixture representation of Linnik law in terms of another: If $0 < \beta < \alpha \leq 2$, then there is a random variable $V_{\alpha,\beta}(\nu) \geq 0$ whose density is

$$g(x, \alpha, \beta) = \frac{\alpha}{\pi} \sin(\frac{\pi\beta}{\alpha}) \frac{x^{\beta-1}}{1 + x^{2\beta} + 2x^{\beta} \cos(\frac{\pi\beta}{\alpha})}$$

such that $Y = X V_{\alpha,\beta}(\nu)$ where $X \stackrel{d}{=} AL(\alpha)$; $Y \stackrel{d}{=} AL(\beta)$. Jacques et al. (1999) proved that the generalized Linnik laws belong to the Paretian family. They have shown that $\lim_{n\to\infty} x^{\alpha} P[X > x] = \frac{\nu}{\pi} \Gamma(\alpha) \sin(\frac{\pi\alpha}{2})$, where X has the Linnik distribution. They also discussed the estimation of parameters of Linnik distribution. Kozubowski (2000) discussed the fractional moment estimation of Linnik parameters. Jayakumar et al. (1995) generalized the Laplace processes of Lawrance (1978) and Dewald and Lewis (1985).

Gaver and Lewis (1980) derived the exponential solution of first order autoregressive equation $X_n = \rho X_{n-1} + \eta_n, n = 0, \pm 1, \pm 2, \ldots$, where $\{\eta_n\}$ is a sequence of independently and identically distributed random variables when $0 \leq \rho < 1$. Lin (1994) proved the characterizations of the Laplace and related distributions via geometric compounding. It is well known that the concept of geometric compounding is related to the rarefaction of renewal processes (Rényi 1956) and to damage models (Rao and Rubin 1964; see also Galambos and Kotz 1978, p. 95).

In Sect. 2 of this paper, we describe some of the properties of generalized Mittag-Leffler distributions. In Sect. 3, we develop an autoregressive process of order one (AR (1)) with generalized Mittag-Leffler distribution. In Sect. 4, geometric generalized Mittag-Leffler distribution is introduced and their properties are discussed. Section 5 deals with autoregressive processes with the above marginal distribution. The extension of these results to k^{th} order case is considered in Sect. 6. In Sect. 7 we describe some applications in various fields.

2 Generalized Mittag-Leffler Distribution

In this Section we introduce a new class of distributions called generalized Mittag-Leffler distribution denoted by GMLD (α, β).

A random variable with support over $(0, \infty)$ is said to follow the generalized Mittag-Leffler distribution with parameters α and β if its Laplace transform is given by

$$\psi(t) = E[e^{-tX}] = (1 + t^{\alpha})^{-\beta}; 0 < \alpha \leq 1, \beta > 0. \tag{1}$$

The cumulative distribution function (c.d.f.) corresponding to (1) is given by

$$F_{\alpha,\beta}(x) = P[X \leq x] = \sum_{k=0}^{\infty} \frac{(-1)^k \Gamma(\beta + k) x^{\alpha(\beta+k)}}{k! \Gamma(\beta) \Gamma(1 + \alpha(\beta + k))} \tag{2}$$

It easily follows that when $\beta = 1$, we get Pillai's Mittag-Leffler distribution (see Pillai 1990). When $\alpha = 1$, we get the gamma distribution. When $\alpha = 1$, $\beta = 1$ we get the exponential distribution. This family may be regarded as the positive counterpart of Pakes generalized Linnik distribution characterized by the characteristic function

$$(1 + |t|^{\alpha})^{-\beta}\,; 0 < \alpha \leq 2, \beta > 0. \tag{3}$$

(see Pakes 1998). Now we shall discuss some properties of generalized Mittag-Leffler distributions.

Theorem 2.1. *Let U_α follows the positive stable distribution with Laplace transform $\psi(t) = exp(-t^{\alpha}), t > 0; 0 < \alpha \leq 1$. Let V_β be a random variable, independent of U_α, and follows a gamma distribution with parameter β and Laplace transform $\phi(t) = \left(\frac{1}{1+t}\right)^{\beta}; \beta > 0$. Then $X_{\alpha,\beta} = U_\alpha V_\beta^{\frac{1}{\alpha}}$ follows generalized Mittag-Leffler distribution GMLD(α, β).*

Proof. The Laplace transform of X is

$$\begin{aligned}
\psi_X(t) &= E\left(e^{-tVU^{\frac{1}{\alpha}}}\right)\\
&= \int_0^{\infty} \psi_U\left(tV^{\frac{1}{\alpha}}\right) dF(v)\\
&= \int_0^{\infty} e^{-t^{\alpha}V} dF(v)\\
&= \left(\frac{1}{1+t^{\alpha}}\right)^{\beta}
\end{aligned}$$

Theorem 2.2. *The p.d.f. of $X_{\alpha,\beta}$ is a mixture of gamma densities.*

Proof. Using Theorem 2.1, we have the c.d.f. of $X_{\alpha,\beta}$ as

$$F_{\alpha,\beta}(x) = \int_0^{\infty} S_{\alpha,v}(x) dF_{\beta}(v) \tag{4}$$

where $S_{\alpha,v}(x)$ is the c.d.f. of a distribution with Laplace transform $exp(-vt^{\alpha})$ and $F_\beta(v)$ is the c.d.f. of gamma distribution with p.d.f. $f(y) = \frac{1}{\Gamma\beta} y^{\beta-1} e^{-y}$.

We can rewrite (4) as

$$F_{\alpha,\beta}(x) = \int_0^{\infty} G_{\beta}(x/y)^{\alpha} dS_{\alpha,1}(y)$$

Hence the p.d.f. of X is

$$\begin{aligned}
f_{\alpha,\beta}(x) &= \frac{d}{dx} F_{\alpha,\beta}(x)\\
&= \int_0^{\infty} \frac{\alpha}{\Gamma\beta} \frac{x^{\alpha\beta-1}}{y^{\alpha\beta}} e^{-(x/y)^{\alpha}} dS_{\alpha,1}(y)
\end{aligned}$$

This shows that $f_{\alpha,\beta}$ is a mixture of generalized gamma densities. For $\beta = 1$, $f_{\alpha,\beta}$ reduces to a mixture of Weibull densities.

Remark 2.1. Lin (1998a,b) has shown that $F_{\alpha,\beta}(x)$ is slowly varying at infinity, for $\alpha \in (0, 1]$ and $\beta > 0$.

Remark 2.2. Pillai (1990) had obtained the fractional moments for $\beta = 1$, as

$$E(X_\alpha^r) = \frac{\Gamma(1 - r/\alpha)\Gamma(1 + r/\alpha)}{\Gamma(1 - r)}; 0 < r < \alpha \leq 1.$$

In a similar manner Lin (1998a,b) obtained the fractional moments for $X_{\alpha,\beta}$ for $0 < \alpha \leq 1$ and $\beta > 0$ as

$$\begin{aligned} E(X_{\alpha,\beta}^r) &= \frac{\Gamma(1 - r/\alpha)\Gamma(\beta + r/\alpha)}{\Gamma(1 - r)\Gamma\beta}; -\alpha\beta < r < \alpha \\ &= \infty \text{ if } r \leq -\alpha\beta \text{ or } r \geq \alpha \end{aligned}$$

Definition 2.1. A probability distribution on $R = (0, \infty)$ is said to be in class L if its Laplace transform $\psi(t)$ satisfies

$$\psi(t) = \psi(kt)\psi_k(t), t \in R, k \in (0, 1), \tag{5}$$

with ψ_k a Laplace transform.

Theorem 2.3. *The GMLD belongs to the class L.*

Proof. The proof follows because of the relation

$$(1 + t^\alpha)^{-\beta} = (1 + k^\alpha t^\alpha)^{-\beta} \left(k^\alpha + (1 - k^\alpha) \frac{1}{1 + t^\alpha} \right)^\beta \tag{6}$$

Theorem 2.4. *The GMLD (α, β) is geometrically infinitely divisible for $0 < \alpha \leq 1$, $0 < \beta \leq 1$.*

Proof. From Pillai and Sandhya (1990), a distribution is g.i.d. if and only if its Laplace transform is of the form $\psi(t) = \frac{1}{1+\phi(t)}$ where $\psi(t)$ has complete monotone derivative (c.m.d.) and $\phi(0) = 0$. Now for GMLD (α, β), the Laplace transform is $\psi(t) = \frac{1}{1+\phi(t)}$ where $\phi(t) = (1+t^\alpha)^\beta - 1$. This has c.m.d. if and only if $0 < \alpha \leq 1$ and $0 < \beta \leq 1$.

Remark 2.3. Also being a mixture of gamma random variables, it is g.i.d. By Pillai and Sandhya (1990), GMLD is a distribution with c.m.d. Hence it is infinitely divisible.

3 First Order Autoregressive Processes with GMLD(α, β) Marginals

Now we shall construct a first order time series model with GMLD marginals. The generalized Mittag-Leffler first order autoregressive process GMLAR(1) is constituted by $\{X_n; n \geq 1\}$ where X_n satisfies the equation

$$X_n = aX_{n-1} + \eta_n; 0 < a < 1, \tag{7}$$

where $\{\eta_n\}$ is a sequence of independently and identically distributed random variables such that X_n is stationary Markovian with generalized Mittag-Leffler marginal distribution. In terms of Laplace transform, (2) can be given as

$$\psi_{X_n}(t) = \psi_{\eta_n}(t)\psi_{X_{n-1}}(at) \tag{8}$$

Assuming stationarity, we have,

$$\begin{aligned}\psi_\eta(t) &= \frac{\psi_X(t)}{\psi_X(at)}\\ &= \frac{(1+a^\alpha t^\alpha)^\beta}{(1+t^\alpha)^\beta}\\ &= \left[\frac{1+(at)^\alpha}{1+t^\alpha}\right]^\beta\\ &= \left[a^\alpha + (1-a^\alpha)\frac{1}{1+t^\alpha}\right]^\beta.\end{aligned} \tag{9}$$

We can regard the innovations $\{\eta_n\}$ as the β-fold convolutions of random variables U_n's such that

$$U_n = \begin{cases} 0 & \text{with probability } a^\alpha \\ M_n & \text{with probability } 1-a^\alpha \end{cases}$$

where M_n's are independently and identically distributed Mittag-Leffler random variables. Mittag-Leffler random variables can be generated easily using the following result. Let E be distributed as exponential with unit mean and let Q be distributed as positive stable with Laplace transform e^{-t^α} then $X = QE^{\frac{1}{\alpha}}$ will be distributed as Mittag-Leffler with Laplace transform $(1+t^\alpha)^{-1}$ (see Kozubowski and Rachev 1999).

Jayakumar et al. (1995) developed an algorithm to generate Linnik random variables. In a similar manner, the GMLAR(1) process can be generated using computers. If $X_0 \stackrel{d}{=} GMLAD(\alpha, \beta)$, then the process is strictly stationary. It is sufficient to verify that $X_n \stackrel{d}{=} GMLAD(\alpha, \beta)$ for every n. An inductive argument can be presented as follows. Suppose $X_{n-1} \stackrel{d}{=} GMLAD(\alpha, \beta)$. Then from (8) we have,

$$\psi_X(t) = \left[\frac{1+(at)^\alpha}{1+t^\alpha}\right]^\beta \left[\frac{1}{1+(at)^\alpha}\right]^\beta = \left[\frac{1}{1+t^\alpha}\right]^\beta$$

Hence the process is strictly stationary and Markovian provided X_0 is distributed as GMLD.

Remark 3.1. If X_0 is distributed arbitrarily, then also the process is asymptotically Markovian with generalized Mittag-Leffler marginal distribution.
Proof.

$$\begin{aligned} X_n &= aX_{n-1} + \eta_n \\ &= a^n X_0 + \sum_{k=0}^{n-1} a^k \eta_{n-k}. \end{aligned}$$

Writing in terms of Laplace transform,

$$\begin{aligned} \psi_{X_n}(t) &= \psi_{X_0}(a^n t) \prod_{k=0}^{n-1} \psi_\eta(a^k t) \\ &= \psi_{X_0}(a^n t) \prod_{k=0}^{n-1} \left[\frac{1+(a^{k+1}t)^\alpha}{1+(a^k t)^\alpha}\right]^\beta \longrightarrow (1+t^\alpha)^{-\beta} \text{as } n \to \infty. \end{aligned}$$

Hence it follows that even if X_0 is arbitrarily distributed, the process is asymptotically stationary Markovian with generalized Mittag-Leffler marginals. We therefore have the following theorem.

Theorem 3.1. *The first order autoregressive process $X_n = aX_{n-1} + \eta_n, a \in (0,1)$ is strictly stationary Markovian with generalized Mittag-Leffler marginal distribution as in* (1) *if and only if the $\{\eta_n\}$ are distributed independently and identically as the β-fold convolution of the random variable $\{U_n\}$ where*

$$U_n = \begin{cases} 0 & \textit{with probability } a^\alpha \\ M_n & \textit{with probability } 1-a^\alpha \end{cases}$$

where $\{M_n\}$ are independently and identically distributed Mittag-Leffler random variables provided $X_0 \overset{d}{=} GMLD(\alpha,\beta)$ and independent of η_n.

Remark 3.2. The model is defined for all values of 'a' such that $a \in (0,1)$. The autocorrelation is given by $\rho(r) = Cor(X_n, X_{n-r}) = a^{|r|}; r = 0, \pm 1, \pm 2, \ldots$

3.1 Distribution of Sums and Bivariate Distribution of (X_n, X_{n+1})

We have,

$$X_{n+j} = a^j X_n + a^{j-1}\eta_{n+1} + a^{j-2}\eta_{n+2} + \ldots + \eta_{n+j}; j = 0,1,2,\ldots$$

Hence

$$\begin{aligned} T_r &= X_n + X_{n+1} + \ldots + X_{n+r-1} \\ &= \sum_{j=0}^{r-1} \left[a^j X_n + a^{j-1}\eta_{n+1} + a^{j-2}\eta_{n+2} + \ldots + \eta_{n+j}\right] \\ &= X_n \left[\frac{1-a^r}{1-a}\right] + \sum_{j=1}^{r-1} \eta_{n+j} \left[\frac{1-a^{r-j}}{1-a}\right]. \end{aligned}$$

Therefore the distribution of the sums T_r is uniquely determined by the Laplace transform

$$\begin{aligned} \psi_{T_r}(t) &= \psi_{X_n}\left(\frac{1-a^r}{1-a}t\right) \prod_{j=1}^{r-1} \psi_\eta \left(\frac{1-a^{r-j}}{1-a}t\right) \\ &= \frac{1}{\left[1+\left(\frac{1-a^r}{1-a}t\right)^\alpha\right]^\beta} \prod_{j=1}^{r-1} \left[a^\alpha + (1-a)^\alpha \frac{1}{\left[1+\left(\frac{1-a^{r-j}}{1-a}t\right)^\alpha\right]^\beta}\right]. \end{aligned}$$

The distribution of T_r can be obtained by inverting the above expression. Next, the joint distribution of contiguous observations (X_n, X_{n+1}) can be given in terms of bivariate Laplace transform as,

$$\begin{aligned} \psi_{X_n,X_{n+1}}(t_1, t_2) &= E[exp(it_1 X_n + it_2 X_{n+1})] \\ &= E[exp(it_1 X_n + it_2(aX_n + \eta_n))] \\ &= E[exp(i(t_1 + at_2)X_n + it_2\eta_{n+1})] \\ &= \psi_{\eta_n}(t_2)\psi_{X_n}(t_1 + at_2) \\ &= \left[\frac{1+(at_2)^\alpha}{1+t_2^\alpha}\right]^\beta \left[\frac{1}{1+(t_1+at_2)^\alpha}\right]^\beta. \end{aligned}$$

Since this expression is not symmetric in t_1 and t_2, it follows that the GMLAR(1) process is not time reversible.

4 Geometric Generalized Mittag-Leffler Distribution

Geometric generalized Mittag-Leffler distribution is introduced and some of its properties are studied.

Definition 4.1. A random variable X on $R = (0, \infty)$ is said to follow geometric generalized Mittag-Leffler distribution and write $X \stackrel{d}{=} GGMLD(\alpha, \beta)$ if it has the Laplace transform

$$\psi(t) = \frac{1}{1+\beta\log(1+t^{\alpha})}, 0 < \alpha \le 2, \beta > 0. \tag{10}$$

Remark 4.1. Geometric Generalized Mittag-Leffler distribution is geometrically infinitely divisible.

Theorem 4.1. *Let $X_1, X_2, \ldots$ are independently and identically distributed Mittag-Leffler random variables and $N(p)$ be geometric with mean $\frac{1}{p}$, $P[N(p) = k]=p(1-p)^{k-1}$, $k = 1, 2, \ldots, 0 < p < 1$. Define $Y = X_1 + X_2 + \ldots + X_{N(p)}$, then $Y \stackrel{d}{=} GGMLD(\alpha, \beta)$.*

Proof. The Laplace transform of Y is

$$\begin{aligned}\psi_Y(t) &= \sum_{k=1}^{\infty}[\psi_X(t)]^k p(1-p)^{k-1} \\ &= \frac{1}{1+\frac{1}{p}\log(1+t^{\alpha})}.\end{aligned}$$

Hence $Y \stackrel{d}{=} GGMLD(\alpha, \frac{1}{p})$.

Theorem 4.2. *Geometric generalized Mittag-Leffler distribution is the limit of geometric sum of $GML(\alpha, \frac{\beta}{n})$ random variables.*

Proof. $(1+t^{\alpha})^{-\beta} = \{1+(1+t^{\alpha})^{\frac{\beta}{n}} - 1\}^{-n}$ is the Laplace transform of a probability distribution since generalized Mittag-Leffler distribution is infinitely divisible. Hence by lemma 3.2 of Pillai (1990),

$$\psi_n(t) = \{1 + n[(1+t^{\alpha})^{\frac{\beta}{n}} - 1]\}^{-n},$$

is the Laplace transform of a geometric sum of independently and identically distribute generalized Mittag-Leffler random variables. Taking limit as $n \to \infty$

$$\begin{aligned}\psi(t) &= \lim_{n\to\infty} \psi_n(t) \\ &= \{1 + \lim_{n\to\infty} n[(1+t^{\alpha})^{\frac{\beta}{n}} - 1]\}^{-1} \\ &= [1+\beta\log(1+t^{\alpha})]^{-1}.\end{aligned}$$

Theorem 4.3. *If W and V are independent random variables such that W has geometric gamma distribution with Laplace transform $\frac{1}{1+\beta log(1+t)}$ and V has a positive stable distribution having Laplace transform $e^{-t^{\alpha}}$, then $W^{\frac{1}{\alpha}} V = U$ where $U \stackrel{d}{=} GGMLD(\alpha, \beta)$.*

Proof. The Laplace transform of U is

$$\begin{aligned}\psi_u(t) &= E(e^{-tW^{\frac{1}{\alpha}}V}) \\ &= \int_0^\infty \psi_V(tW^{\frac{1}{\alpha}})dF(w) \\ &= \int_0^\infty e^{-Wt^\alpha} dF(w) = \frac{1}{1+\beta\log(1+t^\alpha)}.\end{aligned}$$

5 Geometric Generalized Mittag-Leffler Processes

In this section, we develop a first order new autoregressive process with geometric generalized Mittag-Leffler marginals.

Consider an autoregressive structure given by,

$$X_n = \begin{cases} \eta_n & \text{with probability } p \\ X_{n-1} + \eta_n & \text{with probability } (1-p) \end{cases} \tag{11}$$

where $0 < p < 1$. Now we shall construct an AR (1) process with stationary marginal distribution as geometric generalized Mittag-Leffler distribution $GGMLD(\alpha, \beta)$.

Theorem 5.1. *Consider a stationary autoregressive process $\{X_n\}$ with structure given by* (11). *A necessary and sufficient condition that $\{X_n\}$ is stationary Markovian with geometric generalized Mittag-Leffler marginal distribution is that $\{\eta_n\}$ is distributed as geometric Mittag-Leffler provided X_0 is distributed as geometric generalized Mittag-Leffler.*

Proof. Let us denote the Laplace transform of X_n by $\psi_{X_n}(t)$ and that of η_n by $\psi_{\eta_n}(t)$, (11) in terms of Laplace transform becomes

$$\psi_{X_n}(t) = p\psi_{\eta_n}(t) + (1-p)\psi_{X_{n-1}}(t)\psi_{\eta_n}(t).$$

On assuming stationarity, it reduces to the form

$$\psi_X(t) = p\psi_\eta(t) + (1-p)\psi_X(t)\psi_\eta(t).$$

Writing

$$\psi_X(t) = \frac{1}{1+\beta\log(1+t^\alpha)} \quad \text{and solving we get,}$$

$$\psi_\eta(t) = \frac{1}{1+\beta \text{p} \log(1+t^\alpha)}$$

Hence it follows that $\eta_n \stackrel{d}{=} GGMLD(\alpha, p\beta)$.

The converse can be proved by the method of mathematical induction as follows. Now assume that $X_{n-1} \stackrel{d}{=} GGMLD(\alpha, \beta)$. Then

$$\begin{aligned}\psi_{X_{n-1}}(t) &= \psi_{\eta_n}(t)\left[p + (1-p)\psi_{X_{n-2}}(t)\right] \\ &= \frac{1}{1 + p\beta\log(1+t^{\alpha})}\left[p + (1-p)\frac{1}{1+\beta\log(1+t^{\alpha})}\right] \\ &= [1 + \beta\log(1+t^{\alpha})]^{-1}.\end{aligned}$$

Remark 5.1. Note that X_n and η_n belongs to the same family of distributions.

5.1 *The Joint Distribution of X_n and X_{n-1}*

Consider the autoregressive structure given in (11). It can be rewritten as

$$X_n = I_n X_{n-1} + \eta_n,$$

where

$$P[I_n = 0] = 1 - P[I_n = 1] = p, 0 < p < 1.$$

Then the joint Laplace transform of (X_n, X_{n-1}) is given by,

$$\begin{aligned}\psi_{X_n, X_{n-1}}(t_1, t_2) &= E(e^{it_1 X_{n-1} + it_2 X_n}) \\ &= E(e^{it_1 X_{n-1} + it_2(I_n X_{n-1} + \eta_n)}) \\ &= E(e^{(it_1 + it_2 I_n)X_{n-1}})\psi_{\eta_n}(t_2) \\ &= \frac{1}{1+\beta p\log(1+t_2^{\alpha})}\left[\frac{p}{1+\beta\log(1+t_1^{\alpha})}\right. \\ &\quad \left. + \frac{1-p}{1+\beta\log(1+t_1+t_2{}^{\alpha})}\right].\end{aligned}$$

This shows that the process is not time reversible.

6 Generalization to a kth Order Geometric Generalized Mittag-Leffler Autoregressive Processes

Lawrance and Lewis (1982) constructed higher order analogs of the autoregressive (11) with structure as given below,

$$X_n = \begin{cases} \eta_n & \text{with probability } p \\ X_{n-1} + \eta_n & \text{with probability } p_1 \\ X_{n-2} + \eta_n & \text{with probability } p_2 \\ \vdots & \\ X_{n-k} + \eta_n & \text{with probability } p_k \end{cases} \tag{12}$$

where $p_1 + p_2 + \ldots + p_k = 1 - p, 0 \leq p_i, p \leq 1, I = 1, 2, \ldots, k$ and η_n is independent of $\{X_n, X_{n-1}, \ldots\}$.

In terms of Laplace transform, (7) can be given as

$$\psi_{X_n}(t) = p\psi_{\eta_n}(t) + p_1\psi_{X_{n-1}}(t)\psi_{\eta_n}(t) + p_2\psi_{X_{n-2}}(t)\psi_{\eta_n}(t) + \ldots$$
$$+ p_k\psi_{X_{n-k}}(t)\psi_{\eta_n}(t)$$

Assuming stationarity, we get,

$$\psi_{\eta_n}(t) = \frac{\psi_X(t)}{p + (1-p)\psi_X(t)}.$$

This establishes that the results developed in Sect. 5 are valid in this case also. This gives rise to the kth order geometric generalized Mittag-Leffler autoregressive processes.

7 Applications

In thermodynamical or statistical applications, one is interested in mean values of a quantity Z(t). Tsallis (1988) generalized the entropic functional of Boltzmann-Gibbs statistical mechanics that leads to q-exponential distributions. He used the mathematical simplicity of kinetic-type equations to emphasize the natural outcome of this distribution that corresponds exactly to the solution of the kinetic equation of non-linear type; the solution has power-law behavior. Saxena et al. (2004a,b) showed that the fractional generalization of the linear kinetic-type equation also leads to power-law behavior. In both cases, solutions can be expressed in terms of generalized Mittag-Leffler functions. Mittag-Leffler distributions can also be used as waiting-time distributions as well as first-passage time distributions for certain renewal processes. Pillai (1990) developed renewal processes with geometric exponential as waiting time distribution. In a similar manner renewal processes with generalized Mittag-Leffler and geometric generalized Mittag-Leffler waiting times can be constructed.

Acknowledgment The second author is grateful to Kerala State Council for Science, Technology and Environment for the KSCSTE Fellowship under which this research was conducted.

References

Anderson, D.N., Arnold, B.C.: Linnik distributions and processes. Journal of Applied Probability **30** 330–340 (1993)
Dewald, D.N., Lewis, P.A.W.: A new Laplace second order auto regressive time series model-NLAR (2). IEEE Transactions in Information Theory **31** 645–651 (1985)
Devroye, L.: A note on Linnik's distribution. Statistical Probability Letters **9** 305–306 (1990)
Erdélyi, A.: Higher Transcendental Functions. Vol. 3, McGraw Hill, New York (1955)
Fujita, Y.: A generalization of the results of Pillai. Annals of the Institute of Statistical Mathematics **45** 361–365 (1993)
Galambos, J., Kotz, S.: Characterizations of probability distributions. Lecture Notes in Mathematics, Vol. 675, Springer-Verlag, New York (1978)
Gaver, D.P., Lewis, P.A.W.: First-order autoregressive gamma sequences and point processes. Advances in Applied Probability **12** 727–745 (1980)
Haubold, H.J., Mathai, A.M.: The fractional kinetic equation and thermonuclear functions. Astrophysics and Space Science **273** 53–63 (2000)
Jacques, C., Remillard, B., Theodorescu, R.: Estimation of Linnik parameters. Statist. Decisions **17** 213–236 (1999)
Jayakumar, K., Pillai, R.N.: On class L distributions. Journal of the Indian Statistical Association **30** 103–108 (1992)
Jayakumar, K., Pillai, R.N.: The first-order autoregressive Mittag-Leffler process. Journal of Applied Probability **30** 462–466 (1993)
Jayakumar, K., Kalyanaraman, K., Pillai, R.N.: α-Laplace Processes. Mathematics of Computational Modelling **22** 109–116 (1995)
Jayakumar, K., Ajitha, B.K.: On the geometric Mittag-Leffler distributions. Calcutta Statistical Association Bulletin **54** Nos. 215–216, 195–208 (2003)
Jose, K.K., Seetha Lekshmi, V.: On geometric exponential distribution and its applications. Journal of the Indian Statistical Association **37** 51–58 (1997)
Klebanov, L.B., Maniya, G.M., Melamed, I.A.: A problem of Zolotarev and analogs of infinitely divisible and stable distributions in a scheme for summing a random number of random variables. Theory of Probability Applications **24** 791–794 (1984)
Kotz, S., Kozubowski, T.J., Podgorski, K.: The Laplace Distributions and Generalizations. Birkhaeuser, Boston (2001)
Kotz, S., Ostrovskii, I.V.: A mixture representation of the Linnik distribution. Statistical Probability Letters **26** 61–64 (1996)
Kozubowski, T.J., Rachev, S.T.: Univariate geometric stable laws. Journal of Computational Analysis and Applications, preprint (1999)
Kozubowski, T.J.: Fractional moment estimation of Linnik and Mittag-Leffler parameters. Mathematics of Computational Modelling. Special issue: Stable Non-Gaussian Models in Finance and Econometrics **34** 1023–1035 (2000)
Kozubowski, T.J.: Mixture representation of Linnik distribution revisited. Statistical Probability Letters **38** 157–160 (1998)
Lawrance, A.J.: Some autoregressive models for point processes. Colloquia Mathematica Societatis Janos Bolyai **24** Point Processes and Queuing Problems, Hungary, 257–275 (1978)
Lawrance, A.J., Lewis, P.A.W.: A mixed exponential time series model. Management Science **28** 9 1045–1053 (1982)
Lin, G.D.: A note on the characterization of positive Linnik laws. Australian New Zealand Journal of Statistics **43** 17–20 (2001)
Lin, G.D.: A note on the Linnik distributions. Journal of Mathematical Analysis and Applications **217** 701–706 (1998a)
Lin, G.D.: On the Mittag-Leffler distributions. Journal of Statistical Planning Inference **74** 1–9 (1998b)

Lin, G.D.: Characterizations of the Laplace and related distributions via geometric compound. Sankhya **56** 1–9 (1994)

Linnik, Yu.V.: Linear forms and statistical criteria, I, II. Ukrainian Mathematical Zhurnal **5** 207–243 (1962) (English Translations in Mathematical Statistics and Probability **3** 1–40, 41–90, American Mathematical Society, Providence, R.I.)

Pakes, A.G.: Mixture representations for symmetric generalized Linnik laws. Statistical Probability Letters **37** 213–221 (1998)

Pillai, R.N.: Semi-α-Laplace distributions. Communications in Statistical Theoretical Methods **14** 991–1000 (1985)

Pillai, R.N.: On Mittag-Leffler and related distributions. Annals of the Institute of Statistical Mathematics **42** 157–161 (1990)

Pillai, R.N., Sandhya, E.: Distributions with complete monotone derivative and geometric infinite divisibility. Advances in Applied Probability **22** 751–754 (1990)

Pillai, R.N., Jayakumar, K.: Specialized class L property and stationary autoregressive process. Statistical Probability Letters **19** 51–56 (1994)

Pillai, R.N., Jayakumar, K.: Discrete Mittag-Leffler distributions. Statistical Probability Letters **23** 271–274 (1995)

Rao, C.R., Rubin, H.: On characterization of the Poisson distribution, Sankhya, Ser. A, **26** 294–298 (1964)

Rényi, A.: A characterization of the Poisson process. Magyar Tud. Akad. Mat. Kutato Int. Kozl. **1** 519–527 (1956) (in Hungarian). (Translated into English in Selected Papers of Alfred Rényi, Vol.1, Akademiai Kiadó, Budapest, 1976)

Saxena, R.K., Mathai, A.M., Haubold, H.J.: On fractional kinetic equations. Astrophysics and Space Science **282** 281–287 (2002)

Saxena, R.K., Mathai, A.M., Haubold, H.J.: Unified fractional kinetic equation and a fractional diffusion equation. Astrophysics and Space Science **209** 299–310 (2004a)

Saxena, R.K., Mathai, A.M., Haubold, H.J.: On generalized fractional kinetic equations. Physica A **344** 657–664 (2004b)

Seetha Lekshmi, V., Jose, K.K.: Geometric Mittag-Leffler tailed autoregressive processes. Far East Journal of Theoretical Statistics **6** 147–153 (2002)

Seetha Lekshmi, V., Jose, K.K.: Geometric Mittag-Leffler distributions and processes. Journal of Applied Statistical Sciences (accepted for publication) (2003)

Seetha Lekshmi, V., Jose, K.K.: An autoregressive process with geometric α-Laplace marginals. Statistical Papers **45** 337–350 (2004)

Tsallis, C.: Possible generalization of Boltzmann-Gibbs statistics. Journal of Statistical Physics **52** 479–487 (1988)

Weron, K., Kotulski, M.: On the Cole–Cole relaxation function and related Mittag-Leffler distribution. Physica A **232** 180–188 (1996)

Solar Wind Speed Theory and the Nonextensivity of Solar Corona

Jiulin Du and Yanli Song

Abstract The solar corona is a complex system, with nonisothermal plasma and being in the self-gravitating field of the Sun. So the corona plasma is not only a nonequilibrium system but also a nonextensive one. We estimate the parameter describing the degree of nonextensivity of the corona plasma and study the generalization of the solar wind speed theory in the framework of nonextensive statistical mechanics. It is found that, when use Chapman's corona model (1957) as the radial distribution of the temperature in the corona, the nonextensivity reduces the gas pressure outward and thus leads a significant deceleration effect on the radial speed of the solar wind.

Keywords Solar wind theory · Nonextensive statistics

1 Introduction

Solar corona is a complex system. The solar wind flow (coronal outflow) is due to the huge difference in the pressure between the solar corona and the interstellar space. This pressure difference drives plasma outward despite the restraining influence of the solar gravitation. Traditionally, the gas of corona plasma is assumed naturally to follow from the kinetic theory in the Maxwellian sense, where the pressure is taken to be the one of an ideal gas (Hundhausen 1972). However, due to the long-range nature of the gravitating interactions, the gas under the field of solar gravitation behaves *nonextensively*, which would lead to the particles not strictly to follow Maxwellian distribution but, in some situations, to be power-law one (Cranmer 1998; Du 2004a, 2006a, 2007; Lavagno and Quarati 2006). Let us now study the nonextensive effect of the corona plasma in the gravitational field of

J.L. Du (✉) and Y.L. Song
Department of Physics, School of Science, Tianjin University, Tianjin 300072, China
e-mail: jiulindu@yahoo.com.cn

H.J. Haubold and A.M. Mathai (eds.), *Proceedings of the Third UN/ESA/NASA Workshop on the International Heliophysical Year 2007 and Basic Space Science*, Astrophysics and Space Science Proceedings, DOI 10.1007/978-3-642-03325-4_10,

the Sun on the solar wind speed. We will consider this system in terms of the theory of astrophysical fluid dynamics and introduce the nonextensive effect into the solar wind theory.

In the conventional theory for the solar wind speed, the velocity distribution function for the particles has been usually assumed to be the Maxwellian one, which has been incarnated in the theory by the use of the equation of ideal gas as well as the Boltzmann–Gibbs entropy since the basic theory of solar wind was proposed by Parker in 1958. However, observations of space plasmas seldom indicate the presence of ideal Maxwellian velocity distributions. Spacecraft measurements of plasma velocity distributions in the solar wind have revealed that non-Maxwellian distributions are quite common (see, e.g., Cranmer 1998 and the references in). In many situations the distributions appear reasonably Maxwellian at low energies but have a "suprathermal" power-law tail at high energies. This has been well modeled by the κ-distribution (Olbert 1969; Collier et al. 1996), one statistical distribution equivalent to the q-distribution presented now in the framework of nonextensive statistical mechanics (Liu and Du 2008; Liu et al. 2009; Du 2004a, b, 2006a, b, c; Silva et al. 1998) by the power-law form,

$$f(v) = B_q \left[1 - (1 - q)\frac{mv^2}{2kT}\right]^{\frac{1}{1-q}}, \quad (1)$$

It has been found that the nonextensivity of the systems with self-gravitating long-rang interactions, in certain situations, can be described by the above power-law q-distribution. In the solar corona, the κ-like distributions have been proposed to arise from strong nonequilinrium thermodynamic gradients, Fermi acceleration at upwelling convection-zone waves or shocks, and electron-ion runaway in a Dreicerorder electric field (see, e.g., Cranmer 1998 and the references in).

We want to know such a κ-like distribution will produce relatively to the Maxwellian one what effects on the speed of the solar wind rather than to explain the phenomena observed. For this purpose, we first in Sect. 2 simply review the conventional theory for the solar wind speed, and then in Sect. 3 we study the degree of nonextensivity of the solar corona plasma. In Sect. 4, we generalize the theory in the framework of nonextensive statistics and in Sect. 5 we investigate the nonextensive effect on the solar wind speed. Finally in Sect. 6, we give the conclusive remarks.

2 The Conventional Speed Theory for the Solar Wind

As we know that the solar wind plasma is believed to be a nonequlibrium system and the atmosphere of the Sun is assumed to be steady and spherically symmetric. In such situations, the basic fluid dynamical equations in the solar wind theory are given (Zhang 1992; Hundhausen 1972) by the mass equation (the continuity equation),

$$4\pi r^2 \rho v = \text{constant}, \quad (2)$$

the momentum equation,

$$\rho v \frac{dv}{dr} = -\frac{dP}{dr} - \frac{GM}{r^2}\rho + \rho D, \tag{3}$$

and the energy equation,

$$\rho v \frac{ds}{dr} = \frac{Q}{T}, \tag{4}$$

where r is the distance from the center of the Sun; ρ, T, P and s are the density, the temperature, the pressure and the specific entropy, respectively; v is the speed of the fluid, Q is the rate of the internal energy generated by the nonradiative processes (including the heat transfer), and D is the rate of the momentum obtained by the ways except the heat pressure gradient and the gravity. Usually, the pressure of the atmosphere of the Sun takes the form of the equation of state of ideal gas, $P = \rho kT/m$ based on the standard statistical mechanics, with m the mass of the proton and k the Boltzmann constant and the specific entropy takes the form,

$$s = \frac{3}{2}\frac{k}{m}\ln\left(P\rho^{-5/2}\right) + \text{const.} \tag{5}$$

In the literatures that discuss the solar wind theory, one usually makes use of (5) and combine (3) and (4) to derive the differential form of the equations,

$$\frac{d}{dr}\left(\frac{1}{2}v^2 + \frac{5}{2}\frac{kT}{m} - \frac{GM}{r}\right) = \frac{Q}{\rho v} + D. \tag{6}$$

This form can be used to explain why the energy flux brought by the steady convection of the plasma may be changed by the local heating to produce the heat pressure gradient or by directly increasing the momentum, such as the wave pressure gradient. If delete ρ from (2) and (3), we can obtain the so-called continuity momentum equation,

$$\left(v - \frac{v_s^2}{v}\right)\frac{dv}{dr} = \frac{2v_s^2}{r} - \frac{dv_s^2}{dr} - \frac{GM}{r^2} + D, \tag{7}$$

where $v_s = \sqrt{kT/m}$ is the sound speed. Equation (7) tells us that the corona plasma might be accelerated by the combination effects of those terms on the right hand side of (7). Here we will introduce the nonextensivity into the above theory. It might be one possible acceleration effect on the solar wind speed.

3 Temperature, Pressure and Nonextensivity of the Corona

The solar corona is a plasma system far from equilibrium. According to Chapman's corona model (1957), the corona temperature varies dependent on the radial distance from the center of the Sun (Lin 2000; Zhang 1992) as follows,

$$T(r) = T_0 \left(\frac{r_0}{r}\right)^{2/7}, \quad r \geq r_0, \tag{8}$$

where r_0 is a reference distance from the center of the Sun. For example, it may be taken as $r_0 = 1.05R_\odot$, $R_\odot$ is the radius of the Sun. T_0 is the temperature at $r = r_0$. The dependence of pressure on the radial distance is expressed by

$$P(r) = P_0 \exp\left(-\frac{GM\mu}{\Re}\int_{r_0}^{r}\frac{dr'}{r'^2 T(r')}\right), \tag{9}$$

where P_0 is the pressure at $r = r_0$; $G, M, \Re$ and μ are the gravitational constant, the mass of the Sun, the gas constant and the mean molecular weight, respectively.

We now consider the nonextensivity of the solar corona. The nonextensivity is one non-additive property of the nonequlibrium system being an external field with the long-rang inter-particle interactions. We have already known examples of this in the many-body treatment of self-gravitating systems and plasma systems. The degree of the nonextensivity for a nonequilibrium system with the long-range inter-particle interactions such as the gravitational force can be estimated by using the nonextensive parameter q defined in nonextensive statistical mechanics (NSM) in terms of its deviation from unity. The formulation of the parameter q for the self-gravitating system can be given (Du 2004a, 2006a) by

$$1 - q = -\frac{k}{\mu m_H}\frac{dT}{dr} \Big/ \frac{GM}{r^2}, \tag{10}$$

where k is the Boltzmann constant and m_H is the mass of hydrogen atom. The deviation of q from unity is thought to describe the degree of nonextensivity. If substituting (8) into (10), we have clearly understood that *the solar corona is not only a nonequilibrium system but also a nonextensive system*. Therefore, we have to generalize the solar wind speed theory in the framework of NSM so as to take the nonextensive effect into consideration, though the conventional theory has made very extensive applications.

From (8) we can find the temperature gradient in the corona,

$$\frac{dT}{dr} = -\frac{2}{7}T_0\left(\frac{r_0}{r}\right)^{2/7}\frac{1}{r}, \tag{11}$$

and then, make use of (10),

$$1 - q = \frac{2kT_0}{7m}\left(\frac{r_0}{r}\right)^{2/7} \Big/ \frac{GM}{r}, \tag{12}$$

where m is the mass of proton. When we consider the case of $r_0 = 1.05R_\odot$ and $T_0 = 1.8 \times 10^6$K (Zhang 1992), the temperature gradients and the values of 1-q at the different distances from the sun can be obtained from (11) to (12). They are shown in Table 1. We find that the value of $1 - q$ (the degree of nonextensivity) in

Table 1 The values of $1-q$ in the solar corona

r/r_0	T/T_0	dT/dr (K km^{-1})	$1-q$
1	1.00	−0.7037	0.0234
2	0.82	−0.2886	0.0384
4	0.67	−0.1184	0.0631
6	0.60	−0.0703	0.0843
8	0.55	−0.0486	0.1036
10	0.52	−0.0364	0.1200
15	0.46	−0.0216	0.1618
20	0.42	−0.0149	0.1984
30	0.38	−0.0089	0.2667
40	0.35	−0.0061	0.3267
50	0.33	−0.0046	0.3832
60	0.31	−0.0036	0.4365
70	0.297	−0.0030	0.4873
80	0.286	−0.0025	0.5360
90	0.276	−0.0022	0.5831
100	0.268	−0.0019	0.6294

the corona rises with the increase of the radial distance r from the Sun, though the temperature gradient slopes more and more gently. The farther the radial distance is from the Sun, the higher the degree of nonextensivity is, which, as we will see, with the distances far away from the Sun, would have a more significant deceleration effect on the solar wind's speed.

4 A Generalization of the Solar Wind Speed Theory in NSM

The nonextensive statistical mechanics based on Tsallis entropy can be defined by the so-called q-logarithmic function, $\ln_q f$, and q-exponential function, $\exp_q f$, (for example, Tsallis et al. 1998; Lima et al. 2001)

$$\ln_q f = (1-q)^{-1}\left(f^{1-q}-1\right), \quad f>0; \tag{13}$$

$$\exp_q f = [1+(1-q)\,f]^{1/1-q}, \tag{14}$$

with $1+(1-q)\,f > 0$ and $\exp_q f = 0$ otherwise, where q is the nonextensive parameter different from unity. In other words, the deviation of q from unity describes the nonextensive degree of the system under consideration. When $q \to 1$ all the above expressions reproduce those verified by the usual elementary functions, and Tsallis entropy function, $S_q = -k\int f^q \ln_q f\, d\Omega$, reduces to the standard Boltzmann–Gibbs logarithmic one, $S = -k\int f \ln f\, d\Omega$. In this new statistical framework, the pressure of the atmosphere of the Sun can be written (Du 2006a, c) as

$$P_q = C_q \rho kT/m \tag{15}$$

with the q-coefficient $C_q = 2/(5-3q)$, $0 < q < 5/3$. And naturally, the specific entropy can be written by

$$s_q = \frac{3}{2}\frac{k}{m}\ln_q\left(P_q\,\rho^{-5/2}\right) + \text{constant}, \tag{16}$$

After the q-logarithmic function is replaced by (13), one has

$$s_q = \frac{3}{2}\frac{k}{m}\frac{1}{1-q}\left[P_q^{1-q}\rho^{-5(1-q)/2} - 1\right] + \text{constant}. \tag{17}$$

We use (17) and (15) instead of (5) and the equation of state of ideal gas, respectively, then in the new framework, (6) and (7) become

$$\frac{d}{dr}\left(\frac{1}{2}v^2 + \frac{5C_q}{2}\frac{kT}{m} - \frac{GM}{r}\right) = C_q^{2-q}\left(\frac{kT}{m}\right)^{1-q}\rho^{-(5-3q)/2}\frac{Q}{v} + D, \tag{18}$$

and

$$\left(v - C_q\frac{v_s^2}{v}\right)\frac{dv}{dr} = C_q\frac{2v_s^2}{r} - C_q\frac{dv_s^2}{dr} - \frac{GM}{r^2} + D. \tag{19}$$

Thus the nonextensivity of the solar corona plasma has been introduced into the theory of the solar wind. These new equations tell us that the nonextensive parameter q different from unity will play a role in the acceleration of the solar wind speed.

5 The Nonextensive Effect on the Solar Wind Speed

In Parker's theory of solar wind speed (1958), the solar corona is assumed approximately to be isothermal one and the nonextensive effect is neglected because the temperature gradient is thought to be zero. Thus, the fluid dynamical equations are the equation of the conservation of mass, (2), and the equation of the conservation of momentum, (3), while the energy equation, (4), is discarded. In such a case, the dynamical equations for the solar wind speed become directly (7). Namely,

$$\left(v - \frac{v_s^2}{v}\right)\frac{dv}{dr} = \frac{2v_s^2}{r} - \frac{GM}{r^2}. \tag{20}$$

We now consider the nonextensive effects expressed by (10) on the Parker's speed theory. In the present case that the nonextensive effect is taken into consideration, the dynamical equation for the solar wind speed should be replaced by the generalized one, (19). Namely,

$$\left(v - C_q \frac{v_s^2}{v}\right) \frac{dv}{dr} = C_q \frac{2v_s^2}{r} - C_q \frac{dv_s^2}{dr} - \frac{GM}{r^2}. \tag{21}$$

If $q \to 1$ it reduces to the conventional form, (20). In the new dynamical equation, the new critical speed for the solar wind is no longer the sound speed but is $v_c(q) = v_s\sqrt{C_q} = \sqrt{2C_q kT/m}$. The new critical distance $r_c(q)$ can be determined by the equation

$$r_c^2(q) - \frac{2T}{dT/dr} r_c(q) + \frac{2T}{dT/dr} \frac{r_c(1)}{C_q} = 0, \tag{22}$$

where $r_c(1) = r_c(q = 1) = GM/2v_s^2$ is the old critical distance in the Parker' theory when the nonextensive effect is not considered. When r is very large, we consider $dT/dr \to 0$, then we get

$$r_c(q) = \frac{GMm}{4C_q kT} = \frac{r_c(1)}{C_q}. \tag{23}$$

and (21) becomes

$$\left[\left(\frac{v}{v_c(q)}\right)^2 - 1\right] \frac{dv}{v} = 2\left[1 - \frac{r_c(1)}{C_q r}\right] \frac{dr}{r} - 2d \ln v_s, \tag{24}$$

Complete the integral of above equation, we find

$$\left(\frac{v}{v_c(q)}\right)^2 - 2 \ln v = 4 \ln r + \frac{4r_c(1)}{C_q r} - 4 \ln v_s + const. \tag{25}$$

where the integral constant can be determined by using $r = r_c(q), v = v_c(q)$. Thus (25) becomes

$$\left(\frac{v}{v_c(q)}\right)^2 - \ln\left(\frac{v}{v_c(q)}\right)^2 = 4 \ln \frac{r}{r_c(q)} + \frac{4r_c(1)}{C_q}\left(\frac{1}{r} - \frac{1}{r_c(q)}\right) + 1, \tag{26}$$

It determines variations of the radial speed dependent on the radial distance from the Sun. It is clear that all the equations depend explicitly on the nonextensive parameter q. When $q \to 1$, (26) recovers perfectly the speed equation of Parker's solar wind theory under the assumption of isothermal corona (Zhang 1992),

$$\left(\frac{v}{v_c(1)}\right)^2 - \ln\left(\frac{v}{v_c(1)}\right)^2 = 4 \ln \frac{r}{r_c(1)} + \frac{4r_c(1)}{r} - 3, \tag{27}$$

where $v_c(1) = v_s = \sqrt{2kT/m}$.

By the numerical calculations, the radial dependence of the solutions of (26) and (27) are obtained, respectively. They are shown in Fig. 1 The dashed line denotes the solution of (26), being the solar wind speed in the case that the nonextensive

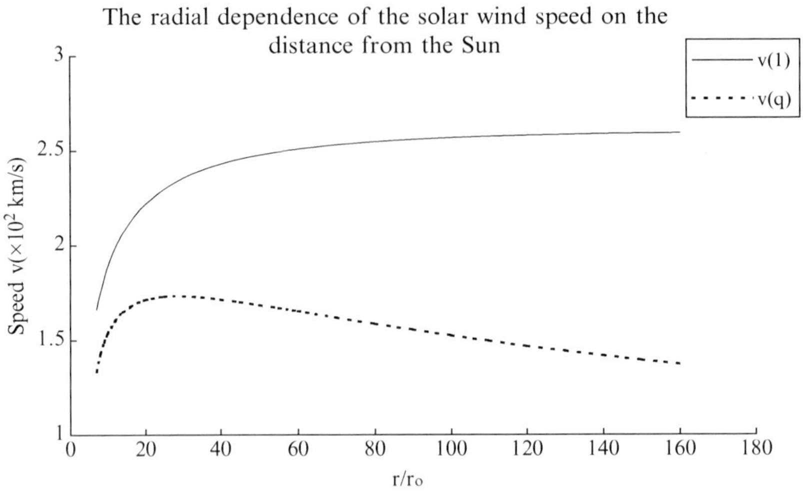

Fig. 1 The radial variation of the solar wind speed dependent on the distance from the center of the Sun. The dashed line denotes the speed in the case that the nonextensive effects are taken into account. The solid line denotes the Parker's speed but considering the radial variation of the temperature

effects are taken into account. The solid line denotes the solution of (27), being the Parker's solar wind speed but considering the radial variation of the temperature. These results clearly show that the nonextensivity of the solar corona has a significant deceleration effect on the radial speed of solar wind. And this effect would be enhanced with the distance far away from the Sun.

From (26) we can derive the speed formula of the solar wind at the distances far away from the Sun. For instance, when r is very large, we have $4r_c(1)/C_q r \to 0$ and, by using (23), $-4r_c(1)/C_q r_c(q) = -4$. Then $-\ln(v/v_c(q))^2$ and -3 are very small as compared with the terms, $(v/v_c(q))^2$ and $4\ln(r/r_c(q))$, and so may be neglected. Thus (26) can be reduced to

$$v^2(q) \approx 4v_c^2(q)\left[\ln\frac{r}{r_c(q)}\right]. \tag{28}$$

It is clear that when $q \to 1$ the speed formula of Parker's solar wind theory for r to be very large, $v(1) \approx 2v_s[\ln r/r_c(1)]^{1/2}$ (Zhang 1992; Hundhausen 1972), can also be recovered correctly from (28). Substitute the related parameters into (28), we find

$$\begin{aligned} v^2(q) &= 4C_q v_s^2\left[\ln\frac{r}{r_c(q)}\right] \\ &= \frac{2}{5-3q}\left[v^2(1) + \frac{8kT}{m}\ln\frac{2}{5-3q}\right]. \end{aligned} \tag{29}$$

By using this formula we can calculate the square of speed of the solar wind far away from the Sun in the case that the nonextensive effect is taken into consideration. For example, the radial speed of solar wind at $r = 100r_0$ can be written as

$$v\,(q = 0.37) = 0.72v\,(1)\left(1 - \frac{4.12 \times 10^{14}}{v^2\,(1)}\right)^{1/2}\ \mathrm{cm\,s^{-1}}. \tag{30}$$

with the nonextensive parameter $q = 0.37$. If take Parker's speed as $v\,(1) = 263\,\mathrm{km\,s^{-1}}$, for example, then we find the new speed to become $v\,(q = 0.37) = 157\,\mathrm{km\,s^{-1}}$. This modification introduced by the nonextensive effects is quite significant.

6 Conclusive Remarks

We have studied the degree of nonextensivity of the solar corona and have introduced the nonextensive effect into the solar wind speed theory. The nonextensivity is shown to be a possible deceleration effect on the radial speed of the solar wind. Our results are concluded as follows:

(a) The solar corona is not only a nonequilibrium system but also a nonextensive system. The degree of nonextensivity in the system rises with the increase of the radial distance from the Sun, though the temperature gradient slopes more and more gently.
(b) The nonextensivity of the solar corona has a significant deceleration effect on the radial speed of solar wind. And this effect will be enhanced with the increase of the distance away from the Sun.
(c) In the new framework of NSM, unlike the case in Parker's theory, the solar wind speed does not rise monotonously with the increase of the distance from the Sun, but rises rapidly with the distance in the regions not very far from the Sun ($< 20R_{\odot}$), being the maximum at about $r = 28r_0$. Further out from the Sun, beyond the maximum, the speed decreases very slowly with the distance away from the Sun.
(d) In Parker's speed equation (26), If we consider the dependence of the corona temperature T on the distance r, as expressed by (7), all the calculated solar wind speeds at the different places are significantly less than those in Parker's solar wind speed theory, under the assumption of isothermal corona.

Why does the nonextensivity have the deceleration effect on the solar wind speed? This question may be answered from the role of the nonextensive effect in the gas pressure of the solar corona. We know that the nonextensive parameter q is determined by (12) and its values are strongly depended on the radial distribution of the temperature in the solar corona. When we take (8) as the temperature model of the corona, the parameter q is always less than unity. The introduction of such a nonextensive effect leads to a decrease of the gas pressure outward (see (15))and thus produces the deceleration of the solar wind speed.

Acknowledgements This work is supported by the National Natural Science Foundation of China under the grant No.10675088 and the "985" Program of TJU of China. We deeply acknowledge H.J. Haubold for continuous encouragement and for his interest in our works on astrophysical applications of statistical mechanics. We thank S. Abe, A.M. Mathai and C. Tsallis for useful discussions.

References

Chapman, S., 1957, Smithson Contr. Astrophys. **2**, 1.
Collier, M.R., Hamilton, D.C., Gloeckler, G., Bochsler, P. and Sheldon, R.B. 1996, Geophys. Res. Lett., **23**, 1191.
Cranmer, S.R., 1998, ApJ. **508**, 925.
Du, J.L., 2004a, Europhys. Lett. **67**, 893.
Du, J.L., 2004b, Phys. Lett. A **329**, 262.
Du, J.L., 2006a, Europhys. Lett. **75**, 861.
Du, J.L., 2006b, Astrophys. Space Sci. **305**, 247.
Du, J.L., 2006c, New Astron. **12**, 60.
Du, J.L., 2007, Astrophys. Space Sci. **312**, 47; cond-mat/0603803.
Hundhausen, A.J., 1972, *Coronal Expansion and Solar Wind*, Berlin: Springer.
Lavagno, A. and Quarati, P., 2006, Astrophys. Space Sci. **305**, 253.
Lima, J.A.S., Silva, R. and Plastino, A.R., 2001, Phys. Rev. Lett., **86**, 2938.
Lin, Y.Z., 2000, *Introduction to Solar Physics*, Beijing: Science Press.
Liu L.Y. and Du, J.L., 2008, Physica A **387**, 4821.
Liu Z.P., Liu L.Y., and Du, J.L., 2009, Phys. Plasmas **16**, 072111.
Olbert, S., 1969, in Physics of the Magnetosphere, ed. Carovillano, R.L., McClay, J.F., and Radoski H.R. Dordrecht: Reidel, 641.
Parker, E.N., 1958, Astrophys. J. **128**, 664.
Silva, R., Plastino, A.R., and Lima, J.A.S., 1998, Phys. Lett. A **249**, 401.
Tsallis, C., Mendes, R.S., and Plastino, A.R., 1998, Physica A **261**, 534.
Zhang, Z.D., 1992, *Solar Physics*, Beijing: Science Press.

Dynamism in the Solar Core

Attila Grandpierre

Abstract Recent results of a mixed shell model heated asymmetrically by transient increases in nuclear burning indicate the transient generation of small 'hot spots' inside the Sun somewhere between 0.1 and 0.2 solar radii (Wolff, 2009). Similar hot bubbles are followed by a nonlinear differential equation system with finite amplitude non-homologous perturbations which is solved in a solar model. Our results show the possibility of a direct connection between the dynamic phenomena of the solar core and the atmospheric activity. Namely, already $\Delta Q_0 \approx 10^{31}$–$10^{37}$ ergs initial heating is enough for a bubble to reach the outer convective zone. Actually, when a hot bubble is enveloped into a magnetic plasmoid, the maintenance of its integrity is enhanced and its surfacing facilitated. Our calculations show that a hot bubble can arrive into subphotospheric regions with $\Delta Q_{\text{final}} \approx 10^{28}$–$10^{34}$ ergs approaching the sound speed $(\sim 10\,\text{km}\,\text{s}^{-1})$. We point out that the developing sonic boom transforms the shock front into accelerated particle beam injected upwards into the top of loop carried out by the hot bubble above its forefront traveling from the solar interior. As a result, a new approach to explain flare energetics had arisen, which, if confirmed, may yield a favorable approach over the best-yet, until now exclusively magnetic flare models. We show that the particle beams generated by energetic deep-origin hot bubbles in the subphotospheric layers have masses, energies, and chemical compositions in the observed range of solar chromospheric and coronal flares. It is shown how the emergence of a hot bubble into subphotospheric regions offers a natural mechanism that can generate both the eruption leading to the flare and the observed coronal magnetic topology for reconnection. We show a tentative list of yet unexplained facts that our model explains, and present a list of predictions for observations, some of which are planned to be realized in the near future.

Keywords Sun · Activity · Interior · Flares · Atmospheric motions · Abundances

---

A. Grandpierre (✉)
Konkoly Observatory, H-1525 Budapest, P. O. Box 67, Hungary
e-mail: grandp@iif.hu

H.J. Haubold and A.M. Mathai (eds.), *Proceedings of the Third UN/ESA/NASA Workshop on the International Heliophysical Year 2007 and Basic Space Science*, Astrophysics and Space Science Proceedings, DOI 10.1007/978-3-642-03325-4_11,

1 Introduction

Recently, a series of new results (Ehrlich, 2007; Wolff, 2009; Hiremath, 2009) indicated that the dynamism of the solar core, which we suggested in our previous papers (Grandpierre, 1996, Paper I.; Grandpierre and Ágoston, 2005; Paper II) can be of interest.

A suggestion made in our Paper I was carried out in more details, and the results led to a new mechanism to explain the Ice Ages of the Earth (Ehrlich, 2007). Moreover, Wolff (2009) pointed out that the g-modes indicate the presence of 'hot spots' we predicted in Papers I and II. He had shown that a whole list of fundamental problems of solar physics have the prospect of being understood if a mixed shell below 0.2 R_{-} permits the excitation of g-modes that form rigidly rotating sets. The overlapping g-modes can represent 3% of the solar luminosity. A mixed shell model heated asymmetrically by transient increases in nuclear burning indicates the transient generation of small 'hot spots' inside the Sun somewhere between 0.1 and 0.2 solar radii. Moreover, Hiremath (2009) had pointed out that the observed quasi-periodicities of solar activity indices in the range of 1–5 years are explained due to perturbations of the strong toroidal field structure and, variation of very long period solar cycle and activity phenomena such as the Maunder and grand minima is explained to be due to coupling of long period poloidal and toroidal MHD oscillations. Since the origin and nature of the solar activity cycle is still a formidable and unsolved problem, it seems to be timely to revise the widespread notion that the solar core is simple. Not so long ago, Bahcall and Pinsonneault (2004, 43) characterized the general view as follows: "The Sun's interior is believed to be in a quiescent state and therefore the relevant physics is simple". But there is a not yet suitably acknowledged reason to think that the solar core has a definite dynamic nature; namely, that the solar core is a magnetized plasma. This fundamental fact escaped attention due to another popular myth considering the Sun as being merely a "luminous ball of gas" (Ridpath 1997, 450). But it is easy to see that the Sun is not a "ball of gas", since it consists largely from ionized particles, and it is penetrated by a magnetic field (as it is shown below). Therefore, the Sun is a vast plasma system coupled to time varying rotation and activity. The Sun can be regarded as a laboratory in which all the four fundamental interactions: electromagnetism, gravitation, weak, and strong interactions, are coupled to each other in vast dimensions. The fact that the solar interior is in a plasma state has a much far reaching significance than it is generally realized nowadays. The electromagnetic interaction is 39 orders of magnitude stronger than the gravitational one; correspondingly, it is enormously richer in nonlinear interactions. Therefore, even if the charges largely balance each other, the remaining small unbalanced forces may dominate behavior. This is the reason why plasma systems show collective behavior. Plasma systems show an enormous variety of dynamic phenomena: being highly electrically conductive, they respond to magnetic fields. Magnetic fields show a tendency to instabilities, have a complex and time varying spatial structure, generating various time varying current systems, filaments, sheets, and jets, manifesting extremely rich behavior. Goossens (2003, 1) emphasized: Plasmas are extremely complicated systems fundamentally different from classic neutral gases, especially when there is a magnetic field present. We will show below by quantitative estimations that magnetic fields play an important role in the plasma of the solar core.

In contrast to a simplified, fictitious plasma of some theoretical approaches, Alfvén (1968) had shown that real plasma show much more complicated behavior. (1) Quite generally a magnetized plasma exhibits a large number of *instabilities*. (2) A plasma has a tendency to produce *electrostatic double layers* in which there are strong localized electric fields. Such layers may be stable, but often they produce oscillations. (3) If a current flows through an electrostatic double layer (which is often produced by the current itself), *the layer may cut off the current*. This means that the voltage over the double layer may reach any value necessary to break the circuit (in the laboratory, say 10^5 or 10^6 V; in the magnetosphere, 10^4–10^5 V; in solar flares, even 10^{10} V). The plasma "explodes", and a high-vacuum region is produced. (4) Currents parallel to a magnetic field (or still more in absence of magnetic fields) have a tendency to pinch; i.e., to *concentrate into filaments* and not flow homogeneously. The inevitable conclusion from phenomena (1)–(4) above is that *homogeneous models are often inapplicable*. Nature does not always have a*horror vacui* but sometimes a *horror homogeneitatis*. In contrast to gases, a plasma, particularly a fully ionized magnetized plasma, is a medium with basically different properties. Typically it is strongly inhomogeneous and consists of a network of filaments produced by line currents and surfaces of discontinuity. These are sometimes due to current sheaths and, sometimes, to electrostatic double layers (Alfven and Arrhenius 1976, Chap. 15).

Regarding the electric aspect of the problem, we can realize that electric instabilities, e.g., plasma microinstabilities generally lead to development of current filaments. In general, the plasma can support electric currents (Goossens 2003, 1). As a rule, the development of instability is accompanied by an increase in the electric field strength, which can attain large values. Consequently, even in the absence of intense external fields, relatively strong fields can still occur spontaneously in plasma due to the growth of instability (Tsytovich 1970, 1). In a highly conductive plasma electric instabilities, especially in changing magnetic fields, develop through complex, nonlinear effects in a way that necessarily lead to a strong amplification of electric currents. Highly amplified electric currents involve extremely strong local heating in the simultaneous presence of inhomogeneous electromagnetic fields and high densities. Recently, Chang et al. (2003) pointed out that the basic MHD equations admit fluctuations to develop that can generate fluctuation-induced nonlinear instabilities reconfiguring the topologies of the magnetic fields.

Apparently, one of the most important keys to the dynamic processes of the solar core lies in the magnetic field. Recently, Gough and McIntyre (1998) had shown that the maintenance of vertical and horizontal shears characterizing the tachocline require their confinement by an underlining magnetic field having a strength of $\sim$1 G just beneath the tachocline. They have argued also that a nonzero interior poloidal field B_i is necessary to explain the observed closely uniform rotation of the radiative zone, and estimated that $B_i \approx 10^3$ G well below the top of the radiative zone if the magnetic field B_i deep in the radiative interior is the remnant of a primordial field.

Friedland and Gruzinov (2004) had shown by solar model calculations that there are many modes of the toroidal field with lifetimes long enough to survive until today. Therefore, the toroidal field in the radiative zone of the Sun can, in principle, have complex structure. The strength of toroidal fields, being entirely confined to the radiative zone, is not subject of the above bound. Friedland and Gruzinov (2004) gave their lowest upper bound to be $B_{FG} \sim 2.1 * 10^6$ G.

It was Hiremath (2005) who determined the poloidal and toroidal parts of the magnetic field profiles in the radiative zone as a function of distance from the solar centre, following the calculations of Hiremath and Gokhale (1995) who used the information of angular velocity as inferred from helioseismology and solved self-consistently the axisymmetric and incompressible MHD equations. Hiremath (2005) found toroidal field strengths up to 10^4 G.

These numerical values for the strength of the magnetic field allow us to estimate the active role of magnetic field in amplifying the dynamism of the solar core's plasma. The criterion for neglecting magnetic effects in the treatment of a problem in gas dynamics is that the Lundquist parameter $L_u = (4\pi)^{1/2} \sigma B l_c c^{-2} \rho^{-1/2}$ (measuring the ratio of the magnetic diffusion time to the Alfven travel time), where σ is the electric conductivity in e.s.u., B is the strength of the magnetic field in Gauss, l_c is a characteristic length of the plasma in centimeter, ρ is the mass density in gcm^{-3}, and c is the speed of light), is much less than unity, $L_u << 1$ (Alfven and Arrhenius 1976, Chap. 15). Now for the solar core $\sigma \approx 10^{17}$ e.s.u., $B \approx 2*10^{-3}$–$2*10^6$ G, $l_c \approx 10^{10}$ cm, $\rho \approx 10^2\,\mathrm{g\,cm}^{-3}$, and so $L_u \approx 7*10^{2-11}$, therefore $L_u >> 1$. This means that plasma effects may play a dominant role in the dynamism of the solar core. For the toroidal field strength of the solar core obtained by Hiremath (2005) $B_H \approx 10^4$ G, we obtain $L_u \approx 3*10^9 >> 1$; for Friedland and Gruzinov (2004) lowest upper bound $B_{FG} \sim 2.1*10^6$ G, the result is $L_u \approx 7*10^{11} >> 1$.

Gervino et al. (2001) determined that the plasma parameter $\Gamma (= e^2/R_D kT$; here 'e' is the charge of the electron, R_D is the Debye screening length, 'k' the Boltzmann constant, 'T' is the temperature of the plasma; Γ measures the mean Coulomb energy potential to the thermal kinetic energy in the solar interior is $\Gamma \approx 0.1$, therefore long-range many-body interactions and memory effects play a significant role. Moreover, they pointed out that the plasma frequency, having a value of 3–$6*10^{17}\,\mathrm{sec}^{-1}$, is of the same order like the collisional frequency; and the screening radius is of the order of the interparticle distance. In addition to many-body collisional effects, electric microfields are present, modifying the usual Boltzmann kinetics. These estimations indicate that highly nonlinear and complex plasma effects may actually and directly play a dominant role in the dynamism of the solar core.

Instead of being a fast rotator as expected, recent helioseismic observations had shown that the solar core rotates almost rigidly. This means that the spin down of the solar core proceeds continuously even nowadays. Hiremath (2001) had shown that differential rotation at the base of the convective zone is more likely than uniform rotation. Therefore it is plausible to allow a small rate of radial differential rotation in the solar core, e.g., $w_1 \sim 10^{-3}\, w_0$ of the rate present in the convective zone. Such a differential rotation generates dissipative processes and, certainly, magnetic instabilities. Certainly, there have to be a coupling between the solar radiative interior and the convective zone, especially if a magnetic field connects these regions. In the case of the Earth, the core-mantle coupling is actually an important trigger contributing to earthquake occurrence (Wang et al. 2004). Therefore, one can consider it plausible that sporadic, singular, localized energy liberation processes, similar to earthquakes, also occur in the solar core. Now Gough and McIntyre (1998) as well as Hiremath (2005) pointed out that the coupling poloidal field has strength in the magnitude of $B_p \approx 1\,G$, a definite value enough to look after measurable consequences. This result makes it plausible to assume that magnetic coupling between the solar core and outer regions generates local energy release in the solar core.

2 Estimations of Local Heating: Spatial and Temporal Scales

Now let us estimate some of the possible effects of magnetic instabilities present in the radiative core of the Sun! Hiremath's equation (Hiremath 2005, 4) offers an estimation for the maximum size of the magnetic instabilities L. Based on simple dimensional analysis and assuming that meridional velocity is very small compared to either the poloidal (B_p) and toroidal (B_t) field strengths or angular velocity, and assuming that both the poloidal and toroidal parts have large diffusion time scales, one can get a relation between differential rotation term w_1 and those parameters (B_p, B_t, η, L) as follows:

$$w_1 \sim \left(\eta/L^2\right) * \left(B_t/B_p\right)$$

where η is the magnetic diffusivity and L is the length scale. By taking the values computed by Hiremath (2005) of B_p and B_t in the radiative zone and for a weak differential rotation rate in the solar core $w_1 \approx 10^{-3}\,w_0$, instabilities may not affect the length scales larger than 100 km. Since the magnetic field structure which Hiremath (2005) proposed is a large-scale one, we don't expect any instability on larger scales. At the same time, these estimations had shown that one has to admit that on the length scales of $L < 100\,\text{km}$ in the radiative zone, plasma instabilities may exist.

The timescale of the simple magnetohydrodynamic Tayler instability (Tayler 1973; Spruit 2002) is very short, of the order of hours and days (Tayler 1973; Goossens and Tayler 1980). It was shown that the most unstable perturbations have a very small wavelength (Goossens and Veugelen 1978). Zirin (1988, 48) argued that the rate of growth of magnetic instabilities is given by $\tau_{magn} \approx L/v_A$, where L is the characteristic spatial scale, and 'v_A' is the Alfven velocity $v_A = B/\,(4\pi\rho)^{1/2}$. For example, with $L = 10^7\,\text{cm}$ and $B_H = 10^4\,G$ one obtains $v_A = 5{*}10^4\,\text{cm}\,\text{s}^{-1}$, and so $\tau_{magn} = 200\,\text{s}$.

Let us estimate the heating energy available by magnetic reconnection. With $w_1 \approx 10^{-3}\,w_0$ and $B_H = 10^4\,G$, when taking a linear size for a spherical region $L = 10^7\,\text{cm}$, one obtains for the local heating a value around $\Delta Q_0\,(\text{magn}) \approx 2{*}10^{28}\,\text{ergs}$; and with $B_{FG} = 2.1{*}10^6\,G$, $L = 10^7\,\text{cm}$, $\Delta Q_0\,(\text{magn}) \approx 7{*}10^{32}\,\text{ergs}$. The calculations of Grandpierre and Ágoston (2005)

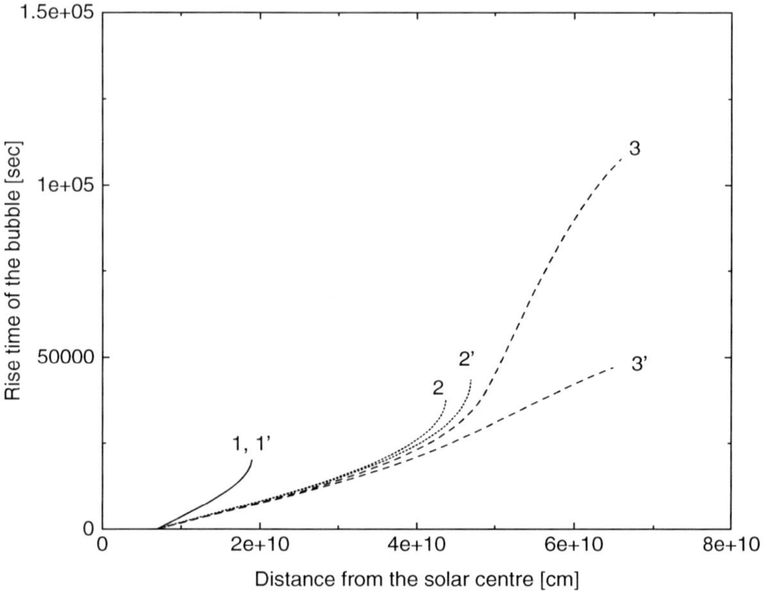

Fig. 1 The rise of bubbles with different initial temperatures, for initial bubble size $R_0 = 5 \times 10^6$ cm, $T_0\left[10^7 K\right] = 1.74$, 5.0, and 9.0, with indices 1, 2 and 3, respectively corresponding to small bubble densities and $\tau_{diff} = \kappa\rho\rho_S C_P R^2/\left(16\sigma\left(T^3 - T_S^3\right)\right)$, while indices 1', 2' and 3' refer to the cases with $\tau_{diff} = \kappa\rho_S^2 C_P R^2/\left(16\sigma\left(T^3 - T_S^3\right)\right)$, corresponding to compressed bubbles and larger bubble densities

had shown that already a heating $\Delta Q_0 \approx 10^{27}$ ergs can be enough to generate a buoyant force which drive the heated region upwards so that the so-formed bubble can make a distance more than its linear size. Depending on the concentration of the heating energy, our calculations indicated that $\Delta Q_0 \approx 10^{32}$ ergs can be enough to drive the bubble upwards to make a significant portion of the solar radius (Figs. 1 and 2, Grandpierre and Ágoston 2005). We note that we consider a plasma system in a strong magnetic field. Magnetic energy dissipation may be expected to occur in a filamentary, highly concentrated form, around narrow current channels. Therefore we can think that reconnection favors highly localized regions to heat.

With a density $\rho \approx 10^2\,\mathrm{g\,cm^{-3}}$ characteristic to the inner solar core, this latter amount of heating energy may lead in the volume with a linear scale $L = 10^7$ cm, to a heating $\Delta T \approx 10$ K. It is easy to see that when $\Delta Q_0 \approx 10^{27}$ ergs heat a smaller, e.g., $L \approx 10^5$ cm region, the arising relative temperature surplus will be $\Delta T \approx 10^7$ K, a value large enough leading to the formation of a hot bubble traveling a path much larger than its diameter. Without any concentration of dissipation, for a certain relative temperature surplus, e.g., when $(\Delta T)_0/T_S \approx 0.1$ and $R \approx 10^6$ cm, we will need an initial heating $\Delta Q_0 \approx 10^{35}$ ergs. More generally, for $\rho \approx 1$–$10^2\,\mathrm{g\,cm^{-3}}$, $R \approx 10^5$–10^7 cm, we will need $\Delta Q_0 \approx 10^{32}$–$10^{38}$ ergs for $(\Delta T)_0/T_S \approx 0.1$ that

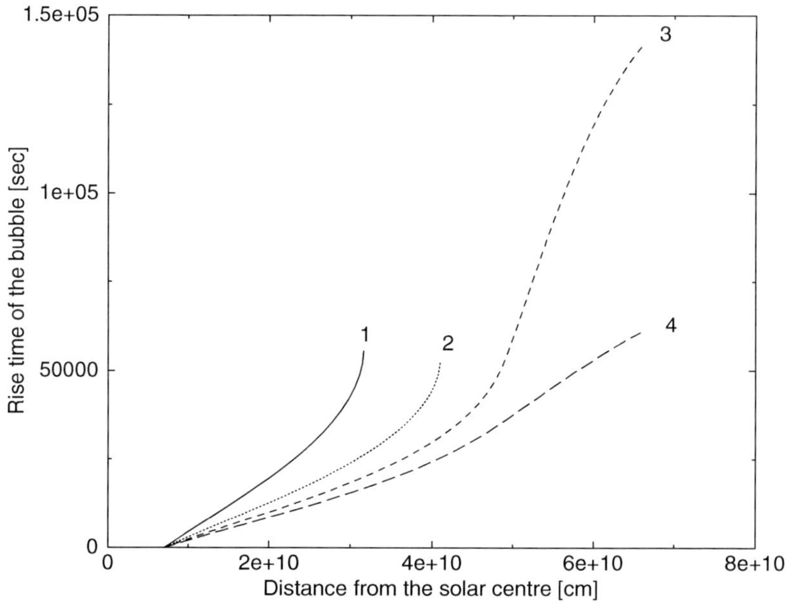

Fig. 2 The rise of bubbles with different initial sizes, for $T_0 = 9.0 \times 10^7\,K$ and $\tau_{diff} = \kappa\rho_S^2 C_P R^2/\left(16\sigma\left(T^3 - T_S^3\right)\right)$. Indices 1, 2, 3 and 4 refers to $R_0 = 10^6$ cm, $R_0 = 2 \times 10^6$ cm, $R_0 = 3 \times 10^6$ cm, $R_0 = 4 \times 10^6$ cm, respectively

Table 1 Available amplitude of heating and initial energy surplus with the different heating mechanisms

Heating mechanisms	Amplitude of heating	Initial temperature surplus
ΔQ_0(magnetic)	$2 * 10^{28}$–$7 * 10^{32}$ ergs	ΔT_0 (magn) $/T_S \sim 10^{-11}$ to > 10
ΔQ_0(rotational)	$2 * 10^{35}$ ergs	ΔT_0(rot)$/T_S \sim 10^{-3}$ to > 10
ΔQ_0(resonant heating)	10^{40} ergs	ΔT_0 (res) $/T_S > 10$

is probably necessary for successful surfacing of the bubble. In Table 1, we present the range of relative heating available by magnetic reconnection.

We emphasize that the basic point of our consideration is that there have to exist some kind of a singular heated region of the solar core. We point out that the lawful development of a singular hot bubble within the solar core has a principal significance. If a certain temperature surplus is present locally, nonlinear couplings within the extremely complex plasma conditions present in the solar core will certainly lead, from time to time, in dependence on local conditions, to amplification of this perturbation to values that are higher by orders of magnitude. Therefore, the estimations presented above merely illustrate situations typically developing in certain localized regions of the solar core – and the numerical calculations below will determine the conditions within which the perturbations will be amplified to observable consequences.

Besides the energies dissipated in magnetic instabilities, rotational energy dissipation is also indicated to become concentrated into small heated regions. In an inhomogeneous body penetrated by a magnetic field, rotational energy dissipation is generally manifested in intermittent local events. During the last $4.6*10^9$ years the solar core has been spun down from a 50 times higher value at the zero-age main sequence (Charbonneau and MacGregor 1992) $E_{rot,0} \approx 10^{45}$ ergs to the present one $E_{rot,present} \approx 2.4*10^{42}$ ergs (Allen 1963, 161). From Fig. 2a of Charbonneau and MacGregor (1992) one can read that the present rate of solar spin-down corresponds to $(\Delta E_{rot}/\Delta t)_{present} \approx 2*10^{34}$ ergs year^{-1}. It is a general view that the dissipation of rotational energy is used to drive the dynamo, and, in general, solar activity in the solar envelope. We note that the main part of the rotational energy is dissipated in the spin-down of the solar core. Now since inhomogeneous electromagnetic field is indicated in the solar core, it seems plausible to assume that ΔE_{rot} is dissipated intermittently and highly localized in the solar core, in a form suitable to drive activity phenomena. We can obtain estimation for the possible amount of heating on the basis that seven "hot spots" was observed during five solar cycles (Bai 2003), therefore the average rate of formation of hot spots is cca. 0.1 year^{-1}. If we identify the formation of a hot spot with a rotational dissipation event, we obtain that the rotational heating has a magnitude ΔQ_0 (rot) $\approx 2*10^{35}$ ergs.

The presence of local heated regions in the solar core is also suggested by Burgess et al. (2003). They indicated the presence of density fluctuations in the deep solar core as a result of a resonant process similar to coronal heating, and had shown that the energy that is transferred from the helioseismic gravity modes into magnetic Alfven modes with density fluctuations leads to strong local heating. They pointed out that the measured spectrum of helioseismic waves does not rule out density variations with amplitudes as large as 10% on scales close to $L \sim 100$ km (again, the same spatial scale limit as that of magnetic instabilities). It is easy to show that such a density variation involves a dissipated heating energy ΔQ_0 (resonant heating) $\approx 10^{40}$ ergs. Now let us see a few timely words about the gravity modes, the necessary input to Burgess's model.

Recently, Garcia et al. (2004) pointed out that some patterns are detected by Turck-Chieze et al. (2004) using GOLF data during the last solar minimum that can be interpreted in terms of gravity modes. The rotation profile is now clearly established down to the limit of the core. In order to progress toward the core and reduce the uncertainties in the radiative region, gravity modes should be measured. Today we cannot arrive at a firmly established conclusion, but we cannot exclude the possible detection of several components of gravity mode candidates. The research of gravity modes detection in the solar core will continue. Turck-Chieze et al. (2004) are confident of making conclusions on the existence of gravity modes in the observations before the end of the life of SOHO in 2007. The possibility that gravity modes will become detectable in the near future, and they may show traces of dynamism, makes it an urgent task to consider theories and models predicting the dynamism of the solar core. In this paper, we developed detailed numerical calculations showing that dynamism of the solar core proceeds in a yet overlooked manner: through singular, individual events like a hot bubble. Stability considerations of the solar core were made only for the shells of the solar core.

Table 2 The time-scales with $R_0 = 3 * 10^6$ cm and different T_0. The bubble rises from r = 0.1 R_{Sun}. The case with $\tau_{diff} = \kappa \rho \rho_S C_P R^2/[16\sigma(T^3 - T_S^3)]$

T_0[in10^7K]	τ_{exp}	τ_{diff}	τ_{cool}	τ_{nucl}	τ_{rise}
1.74	$5.7 * 10^4$	$7.8 * 10^6$	$5.6 * 10^4$	$1.1 * 10^{14}$	$2.3 * 10^4$
5.0	$3.3 * 10^4$	$6.6 * 10^4$	$2.2 * 10^4$	$6.9 * 10^7$	$1.3 * 10^4$
9.0	$3.0 * 10^4$	$6.2 * 10^3$	$5.1 * 10^3$	$2.6 * 10^5$	$1.2 * 10^4$
20.0	$2.9 * 10^4$	$2.5 * 10^2$	$2.5 * 10^2$	$7.8 * 10^1$	$1.2 * 10^4$

Paterno et al. (1997) reconsidered nonradial thermal instabilities in the solar core for internal, infinitesimal, homologous, i.e., shellular perturbations. They have found that the solar core is stable against such perturbations. Remarkably, already on the basis of their result we may conjecture that the solar core is close to instability for finite amplitude nonradial perturbations. This circumstance is due to the fact that the heating timescale they obtained – for homologous perturbations produced by nuclear heating: $\tau_{growth} \sim 4 \times 10^6$ years is only slightly higher than the cooling one, arising from radiative diffusion: $t_{decay} \sim 7 \times 10^5$ years (see their Table 2; at the solar centre). In this paper, we point out that perfect spherical symmetry is impossible in the real Sun. The consequences of significant singular deviations from spherical symmetry seem to be overlooked in the context of instabilities in the solar core. For a highly localized, singular heating, stability analysis was not performed yet, and carrying out such an investigation may be regarded as an important task. We note here that the necessity of a dynamic solar core model is already indicated by many independent theoretical and observational arguments (Grandpierre 1990, 1996), and a trend towards the dynamical representation of the stars is noted (Turck-Chieze 2001).

Besides the above theoretical arguments underpinning the sporadic localization of energy liberation in the plasma of the solar core, we also may have some observational support indicating the presence of heated regions and flare-like phenomena in the solar radiative interior. There are well-founded reasons telling that the observations of Toutain and Kosovichev (2001) and Chaplin et al. (2003) may be signs of flare-like events in the deep solar core. Chaplin et al. (2003) found an anomalous event at late March 1998 supplying additional energy to solar activity and low-l solar p-modes. This event raised the velocity power (V^2, which is directly proportional to the total energy of a mode) by 22% above the zero change level; the predicted value for this epoch in the cycle, however, is of the order of $\sim (-5)$ %. By our best knowledge, similar energy enhancements of p-modes are observed until now only in relation to flares (Haber et al. 1988; Kosovichev and Zharkova 1998). Chaplin et al. (2003) noted that the increase of energy supply is coincident in time with the southern hemisphere onset of cycle 23, with a major emission of particles and the appearance of major surface activity on this hemisphere. Remarkably, Benevolenskaya (1999) had shown that the transition from cycle 22 to cycle 23 clustered in the very same fixed longitudinal regions. Recognizing that such activity enhancements are usually related to active regions with especially high flare activity, and that this event occurred well below the photosphere, one may assume that the increase of energy supply is related to a certain localized event somewhere in the solar interior.

Because this event is energetic and localized, one may apply the term "flare-like event". Now it is an interesting point raised by Bai (2003) – see below – that the hot spots had to form in regions independent of toroidal magnetic flux tubes, in the radiative zone. Therefore, the formation of hot spots should be related to flare-like events are indicated to occur in the radiative zone by the following observations.

Bai (2002) paid attention to the fact that solar flares from the southern hemisphere during cycle 23 are found to be concentrated in a pair of hot spots rotating with a synodic period of 28.2 days, slightly surpassing the range of rotational periods observed both on the surface and in the convective zone in the latitude zone extending from $-35°$ to $35°$, 26–28 days. Moreover, Bai (2003) has been found that the hot spots of this double hot-spot system are separated by about 180° in longitude. Many hot-spot systems last for more than one solar cycle, and therefore the mechanism(s) generating them must be independent of the dynamo working in and around the convective zone. Since the toroidal fields are found around the top of the radiative zone, the mechanism(s) generating the hot spots must act below the zone containing the toroidal flux tubes. Taking into account the facts those hot-spot systems set up frequently in a 180° longitudinal separation, and that they have an anomalous rotation rate from 25 to 29 days, one may seem plausible to find the origin of hot spots deep in the solar core. Actually, helioseismic measurements allow such anomalously rotating layers or regions if their spatial scales are less than 100 km. Therefore, the localization of the source of hot spots suggests the presence of localized hot regions deep in the solar core. It seems plausible to allow that the source of hot spots may be related to the heating events which produce the increased energy supply for solar activity and p-modes in March 1998 (Chaplin et al. 2003).

These theoretical and observational results all indicate that the solar core tend to form sporadic localized heated regions. Therefore, it is important to consider the development of finite amplitude local heating in the solar core. In this paper, we show that in the solar radiative interior sporadically generated hot bubbles may travel significant distances towards the surface. We found that the generation of heated regions presents a new, yet not considered type of instability that lends certain dynamism to the solar core which may have a fundamental significance in the origin of solar activity.

In concluding this section, we present a small table (Table 1), summarizing our findings on the amplitudes of local heating, arising from the given mechanisms. While below $T \sim 10^8$ K cooling has two important mechanisms, volume expansion and radiative diffusion, above $T \sim 10^8$ K radiant cooling will be dominant, and therefore our approach is not valid already. Therefore, in Table 1, we indicated large heating simply by $\Delta T_0/T_S > 10$.

These results are confirmed by our detailed numerical calculations presented below. The large values of temperature surplus of heated bubbles illustrate that all the three mechanisms that are indicated to be plausible sources of heating can lead to significant heating. Therefore, it is instructive to carry out numerical simulations, taking into account the most important heating and cooling processes.

The investigation of bubble-like perturbations creates a new situation in comparison to the already considered shellular case. Bubble formation can couple hydrodynamic instabilities to thermal perturbations. Therefore, it is interesting to follow such finite amplitude bubble-like perturbations individually by numerical computations. In this way, we can determine the parameters of the arising hydrodynamic movements, including the distance a heated bubble may travel, and this parameter may be an important indicator of the dynamism of the solar core.

3 On the Real Physics of the Sun and the Basic Equations of Hot Bubbles

We pointed out in the Introduction that the plasma nature of the solar core presents a complication that has a far reaching significance. We may add that solar activity is a macroscopic phenomenon that occurs as a collective phenomenon involving the cooperation of enormous numbers of particles. The problem of how macroscopic phenomena arise from properties of the microscopic constituents of matter is basically a quantum mechanical one (Sewell 1986, 4). The Sun is composed of approximately 10^{56} interacting particles of several species. At a microscopic level, therefore, its properties are governed by the Schrödinger equation for this assembly of particles. However, the Schrödinger equation of the Sun is extremely complicated: indeed its extreme complexity represents an essential part of the physical situation. Due to this extreme complexity, collective quantum fields become dominating over the individual fields of particles.

Actually, the Sun is a system that is extremely more complicated than its constituents. Being much more than the sum of its constituents, the Sun belongs to the most complex systems of the universe, showing an unusually wide range of emergent complex phenomena: solar activity. In the case of the Sun, not only the collective quantum fields become dominating over the individual ones, but also the gravitational and the electromagnetic fields. Moreover, these dominating cooperative collective fields do interact with each other as well as with the nuclear fields and energy production. Therefore, the Sun is extremely more complex than complex systems in the Earth. This extreme complication leads to computational problems. In order to make the problem solvable, one has to find the physically most interesting aspects of this complexity that may offer a simplification leading to a tractable formulation of the problem. The preliminary consideration of the Introduction served calling attention to the problem of highly localized, singular heating in the solar core.

If we restrict ourselves to describe the rise of the bubbles, the extremely complicated problem simplifies to a tractable one that can be described by the Navier-Stokes equation as the equation of motion, the energy equation and the equation of state.

$$\rho(\partial v_i/\partial t + v_k \partial v_i/\partial x_k) = f_i - \partial p/\partial x_i + \eta \partial^2 v_i/\partial x_k \partial v_k \ (i = 1, 2, 3) \qquad (1)$$

where f_i is the i-component of the total force acting per unit mass and p is the pressure. The conservation of energy tells that

$$dU/dt + p/\rho \mathrm{div}\, v = e_N - 1/\rho \mathrm{div}\,(F_R + F_c) + (p/\rho)\, \partial v_k/\partial x^i, \tag{2}$$

where U is the total thermal energy, ε_N is the liberated nuclear energy per unit mass and time, F_R and F_c are the radiative and conductive fluxes.

The equation of state is

$$p = \left(R_g/\mu\right) \rho T \tag{3}$$

where μ is the dimensionless mean molecular weight, and R_g is the gas constant.

4 Basic Estimations for the Case of a Heated Bubble

We formulate the following scenario: a dissipation event heats a local parcel of matter in the solar interior at some depth (at first, we selected $r = 0.1\, R_{Sun}$). We calculated how this initial perturbation generates a heated bubble (or heated region, in the absence of bubble formation) which is already in pressure equilibrium with its surroundings.

A heated bubble is not in hydrostatic equilibrium with its surroundings. In the first approximation of the Navier-Stokes equation, the motion of a heated bubble is determined by the equality of the buoyant ($F_b = Vg\Delta\rho$) and frictional ($F_f = K/2v^2\, S\rho$) forces, where S is the cross section of the bubble, ρ is the density of the bubble, V is its volume, K is the coefficient of turbulent viscosity, ρ_S and $\Delta\rho$ are the density of the surroundings and the density difference between the bubble and its surroundings, and g is the gravitational acceleration. Equating these forces, $v^2\,(K/2)\,(S/V) = g\Delta\rho/\rho_S$. Now assuming pressure equilibrium between the bubble (referred with no index) and its surroundings (referred with index S), $\rho T = \rho_S T_S$, $\Delta\rho/\rho_S = (1 - T_S/T)$. Taking $K = 1$ (Öpik 1958), we obtain for the bubble's velocity $v = (8/3Rg\,(1 - T_S/T))^{1/2}$. With local heating events in the solar core $\Delta T/T_S \approx 0.1$ (from our Table 1) yields $T_S/T \approx 8/9$, $g \approx 2{*}10^5\ \mathrm{cm\,s^{-2}}$, $R = 10^5$–10^6 cm, the rising speed of the bubble is $v \approx 2$–$7{*}10^5\ \mathrm{cm\,s^{-1}}$.

Now we turn to the energy equation. The heated bubble is not in thermal equilibrium with its surroundings. Below $T \approx 10^8$ K the radiation energy and pressure may be neglected compared to the material energy and pressure. The radiation energy must, of course, not be neglected in the flux term. In a co-moving frame, without energy sources, when radiation is the most effective dissipative factor, the energy equation may be simplified to the form

$$\partial U/\partial t = -1/\rho \mathrm{div}\, F_R = -1/\rho \mathrm{div}\,(D_R\, \mathrm{grad}\, E_R)\,, \tag{4}$$

where $E_R = aT^4$ is the radiation energy density, and a is the radiation-density constant. Assuming that one can apply the diffusion approximation $D_R = 1/3\, cl_{ph}$, where c is the speed of light, and $l_{ph} = 1/(\kappa\rho)$ is the mean free path of a photon, κ is a mean absorption coefficient. In the case of spherical symmetry, the corresponding diffusive radiative flux is

$$F_R = -(4ac/3k\rho)\, T^3 \partial T/\partial r. \quad (5)$$

Now returning to the simplified energy (5), with $U = C_p T$, and integrating it to the whole volume of the bubble,

$$C_p\, (\partial T/\partial t)\, V = -(1/\rho)\, 4\pi\, R^2\, F_R. \quad (6)$$

From this equation the thermal adjustment time is estimated, in a linear approximation, writing for $\partial T/\partial t \sim -\Delta T/\tau_{adj}$, and for $\partial T/\partial r \sim \Delta T/R$, following Kippenhahn and Weigert (1990, 44) as:

$$\tau_{adj} = k\rho^2\, C_p\, R^2/\left(16\sigma T^3\right), \quad (7)$$

where $\sigma = ac/4$ is the Stefan-Boltzmann constant ($\sigma = 5.67{*}10^{-5}\,\mathrm{erg\,cm^{-2}\,K^{-4}\,s^{-1}}$).

With typical values, $\kappa = 2\,\mathrm{cm^2\,g^{-1}}$, $\rho = 90\,\mathrm{g\,cm^{-3}}$, $C_p = 2.1{*}10^8\,\mathrm{erg\,K^{-1}\,mole^{-1}}$, $T = 10^8$ K, $R = 10^6$ cm, $\tau_{adj} = 3{*}10^3$ s, while for $T = 10^7$ K, $\tau_{adj} = 4{*}10^6$ s.

To obtain a preliminary picture on the question whether heated bubbles may travel a distance larger than their characteristic sizes, first we determined the relevant timescales of this process. It is a favorable method because it offers a fast and easy way to obtain a first view on the relations between heating and bubble rise.

In order to put this fundamental thermal timescale into the context we are interested in, we define a time-scale for the rise of the bubbles as $\tau_{rise} = l_T/v$, where l_T is the temperature scale height (in the solar core at $r = R_{Sun}/10$, $l_T \sim 1.5{*}10^{10}$ cm). With $v = 1.5{*}10^5$–$1.5{*}10^6\,\mathrm{cm\,s^{-1}}$, $\tau_{rise} \sim 10^5$–10^4 s, respectively. This means that for (at least) moderate heating (when $T/T_S > 1.0001$) the bubble may move so fast that its thermal cooling is slower than the decrease of the temperature of its environment on its path rising towards the surface (for a moment, we ignore the cooling arising from volume expansion; more detailed results are given later on). In such a case, the bubble cannot adjust its temperature to its environment, and the heating may lead to the formation of a bubble and its self-maintaining rise upwards – even if we disregard from any internal energy source.

The timescale of cooling of the bubbles arising from adiabatic volume expansion may be calculated following Gorbatsky (1964). Starting from

$$Q = C_V\, m\, T = 2\,\pi\, R^3\, p,$$
$$(dQ/dt)_{exp} = -p d\left(4/3\pi R^3\right)/dt$$

Table 3 The time-scales when the bubble rises from $r = 0.65\,R_{Sun}$ (a) with $\tau_{diff} = \kappa\rho\rho_S C_P R^2/\left[16\sigma\left(T^3 - T_S^3\right)\right]$, with $R_0 = 5 * 10^5$ cm and (b) $\tau_{diff} = \kappa\rho_S^2 C_P R^2/\left[16\sigma\left(T^3 - T_S^3\right)\right]$, with $R_0 = 2 * 10^6$ cm

T_0[in 10^7K]	τ_{exp}	τ_{diff}	τ_{cool}	τ_{nucl}	τ_{rise}
6.2 (a)	$1.4 * 10^5$	$2.1 * 10^2$	$2.1 * 10^2$	$3.5 * 10^{10}$	$8.7 * 10^4$
6.2 (b)	$2.4 * 10^5$	$3.6 * 10^{-1}$	$3.6 * 10^{-1}$	$3.5 * 10^{10}$	$4.4 * 10^4$

$$2\pi p\left(3R^2\right)(dR/dt)_{exp} + 2\pi R^3 (dp/dt)_{adiab} = -p\pi R^2 (dR/dt)_{exp}$$
$$(dR/dt)_{exp} = -R/5\,(1/p\,(dp/dt))$$
$$\tau_{exp} = -\left(1/5\,(1/p\,(dp/dr))\,v\right)^{-1} = 5\,H_p/v.$$

Around $r = 0.1R_{Sun}$, the pressure scale height $H_p = |1/p\,(dp/dr)\,|^{-1} \sim 7.3{*}10^9$ cm. For $R = 10^5$–10^6 cm and with $v \sim 10^3$–$10^6\ \mathrm{cm\,s^{-1}}$, τ_{exp} is usually in the range of $3{*}10^7$–10^4 s. This is an important result, since it indicates that the rise time $\tau_{rise} \sim 10^4$–10^5 s and the adiabatic expansion timescale $\tau_{exp} \sim 10^4$–$3{*}10^7$ s somewhat overlap. More concretely, as Table 2 indicates, the rise time is lower than the expansion timescale; therefore the bubble may make significant distances, since it rises faster than it cools by adiabatic expansion. Actually, with $v \sim 10^6\ \mathrm{cm\,s^{-1}}$, to make a distance $R_{Sun} \sim 7{*}10^{10}$ cm, the bubble needs $\tau_{final} \sim 10^5$ s.

It is easy to make the calculations given in Tables 2 and 3. We obtained that $\tau_{exp} \approx 5\,H_p/v$, $\tau_{diff} = \kappa\rho^2 C_P R^2/\left(16\sigma\left(T^3 - T_S^3\right)\right)$, $\tau_{cool} = \left(\tau_{exp}^{-1} + \tau_{diff}^{-1}\right)^{-1}$, $\tau_{rise} \approx 1.5{*}10^{10}/v$. We used τ_{diff} instead of τ_{adj} since we took into account the fact that when the heated bubble temperature approaches the temperature of its surroundings, its diffusive radiative flux decreases to zero. We tested these preliminary estimations with detailed numerical calculations, considering that the material heated by the heat wave of radiative diffusion expanding from the bubble is coupled to it (see Gorbatsky 1964).

The local enhancement of nuclear energy liberation needs a heating timescale $\tau_{nucl} = C_p T/\varepsilon\nu$ (Grandpierre 1990) to be shorter than the timescale of the cooling processes $\tau_{exp} = 5\,H_p/v$, $\tau_{adj} = \kappa\rho^2\,C_p\,R^2/\left(16\sigma T^3\right)$, where C_p is the specific heat at constant pressure, T the temperature of the heated region, ε is the rate of energy liberation by nuclear reactions, ν is the exponent in the $\varepsilon \sim T^\nu$ relation, H_p is the pressure scale height, 'v' the velocity of the heated region, κ is a mean absorption coefficient, ρ and R is the density and the radius of the heated region, σ is the Stefan-Boltzmann constant ($\sigma = 5.67{*}10^{-5}\ \mathrm{erg\,cm^{-2}\,K^{-4}\,s^{-1}}$). With typical values ($\kappa \sim 2\,\mathrm{cm^2\,g^{-1}}$, $\rho \sim 100\,\mathrm{g\,cm^{-3}}$, $C_p \sim 3{*}10^8\ \mathrm{erg\,K^{-1}\,mole^{-1}}$, $T \sim 10^7$ K, $R \sim 10^6$ cm, $H_p \sim 7{*}10^9$ cm, $v \sim 10^3$–$10^6\ \mathrm{cm\,s^{-1}}$), $\tau_{adj} \sim 7{*}10^6$ s, $\tau_{exp} \sim 3{*}10^4$–10^7 s, $\tau_{nucl} \sim 10^{16}$ s, while for $T \sim 10^8$ K, $\tau_{adj} \sim 7{*}10^3$ s, $\tau_{exp} \sim 3{*}10^4$–10^7 s, $\tau_{nucl} \sim 1$ s.

This means that when the sporadic and localized energy dissipation processes indicated in our Table 2 heat a small macroscopic region more then tenfold, above 10^8 K, nuclear energy liberation may make the region explosive, since the diffusion of radiation and volume expansion together cannot cool the heated volume effectively on such a short timescale. At such temperatures, the cooling timescales are more than three orders of magnitude larger than that of the nuclear heating; therefore, local thermonuclear runaway will develop. The volume of the heated region may explosively increase until it forms such a heated bubble that, accelerated by the buoyant force, rises up and transports outwards most of the produced surplus energy.

From these estimations we can recognize the remarkable situation that all the four relevant time-scales determining the behavior of the bubbles $\tau_{\rm rise}$, $\tau_{\rm exp}$, $\tau_{\rm adj}$ and $\tau_{\rm nucl}$ are comparable to each other. Therefore, it is important to consider the case by more detailed numerical calculations and determine if there exist suitable conditions for triggering instability.

5 Method of Calculation

We start by picking up a certain determined virtual value for the radius of the bubble R_V and for its initial virtual temperature surplus $n = T_V/T_S$. These virtual values are not physical values but they soon will turn into realistic values by pressure equilibration. At the very first phase of the bubble formation the density of the heated bubble is $\rho_V = \rho_S$, and $T_v = nT_S$, $Q_V = 2\pi\mu_0^{-1} R_g R_V^3 \rho_V T_V$, $p_V = p_0 = np_S$, $n > 1$. Then we determine the parameters of the bubble which underwent pressure equilibration and is already in pressure equilibrium with its environment (denoted with indices "0"), $\rho_0 = \rho_V n^{-3/5}$, $R_0 = R_V n^{-1/5}$, $Q_0 = Q_V n^{-2/5}$, $T_0 = Q_0 (\mu_0/2\pi) \left(R_0{}^3 R_g \rho_0\right)^{-1}$, $m = \rho_0 (4\pi/3) R^3$. At $t = 0$, $m_+ (t = 0) = 0$ (here 'm' and 'm_+' are the initial mass of the heated bubble, and the mass of the volume heated by radiative diffusion of the bubble, respectively). Then we pick up a certain set of time steps, and determine the values of the parameters in the next time step.

We worked with a fourth order Runge-Kutta method to solve the differential equation system. Our calculations differ from such previous ones like Rosenbluth and Bahcall (1973), and Paterno et al. (1997), who worked with $\Delta\rho = 0$, and $\Delta (T\mu) = 0$, since they considered merely homologous, strictly non-radial perturbations. In our calculations, we allowed non-homologous, singular perturbations, non-vanishing only in a highly localized region, without a strict local hydrostatic equilibrium, and so the heated region may have initially pressure surplus, too. After a transient period lasting for a few seconds pressure equilibrium sets up, and the bubble is hotter and less dense than its surroundings, $\Delta p \sim \Delta (T\rho) = 0$, but $\Delta T_0 \neq 0$ and $\Delta\rho_0 \neq 0$.

We solved the differential equation system with a numerical code (a simpler version of the code is described in Grandpierre and Ágoston 2005). We neglected the radiation pressure in all the terms except the diffusive one. Our method may be regarded as working well for our purposes below 10^8 K, since the estimated error in each quantity is smaller than 15%.

6 Results and Discussion

The calculated timescales show us that the characteristic rise time of the bubbles are comparable (or shorter) than their combined cooling time scales, therefore the bubbles are able to rise significant distances in the radiative core. We note that our calculations involved turbulent drag only approximately. It is possible that in reality turbulence may disturb the hot bubbles in such a rate that they become disintegrated before running distances much larger than their diameter. At the same time, the formation of enveloping plasmoid around the bubbles may be favorable for enhancing their lifetimes. Bubbles in water usually survive rise distances much larger than their linear sizes. Hot bubbles enveloped in a plasmoid may also survive large distances.

The obtained results lead to an important conjecture: namely, there exist a yet unexplored type of stellar instability within the solar core and similar stellar radiative interiors. Figure 1 illustrates the effect of density difference. The difference between curves 1 and 1', 2, and 2', 3, and 3' corresponds to a change from $\tau_{diff} = \kappa\rho\ \rho_S C_P R^2/\left(16\,\sigma\left(T^3 - T_S{}^3\right)\right)$ to $\tau_{diff} = \kappa\rho_S{}^2\ C_P R^2/\left(16\,\sigma\left(T^3 - T_S^3\right)\right)$. The mechanism(s) compressing the bubble on its path may be due to the aerodynamic drag and, more importantly, to the tension of environmental deeply rooted magnetic field lines that the bubbles met on their pathway and that the rising bubbles elongate and push at their forefront upwards. Magnetic fields when forming plasmoid structure around the bubble may serve simultaneously as an accelerator agent through the accompanying magnetic buoyancy effect. When the rising bubble does not decrease its density so fast, due to plasmoid confinement, radiation will escape from its surface in a lesser rate, and so the bubble can travel larger distances, as our detailed calculations show.

We found that when heating above a certain energy threshold is present, it can directly initiate from time to time large-amplitude individual motions of heated bubbles that can travel significant distances within the solar body. Our calculations indicated that this threshold is around $\Delta Q_0 \approx 10^{27}$ ergs. Larger bubbles and larger heating may lead to bubbles traveling much larger distances. Figure 2 shows the effect of the initial spatial size of the bubble. Indices 1, 2, 3, and 4 refers to $R_0 = 10^6$ cm, $2{*}10^6$ cm, $3{*}10^6$ cm, and $4{*}10^6$ cm, respectively. Although smaller bubbles make also significant distances, there exist a certain critical range of spatial sizes (corresponding to different amount of initial heating, see Table 1), above which the bubble may reach the surface regions.

Figure 3 shows the evolution of the rising speed of the bubble. The bump around $r = 1.2$–$1.4{*}10^{10}$ cm in the rising speed 'v' is related to the slow increase of radius R and the local maximum of the gravitational acceleration $g_{max} = 2.4{*}10^5\ \mathrm{cm\,s^{-2}}$ at $r{\sim}1.05{*}10^{10}$ cm that is followed by a fast decrease of g since 'v' is proportional to $(Rg)^{1/2}$. Figure 3 shows that for $T < 10^8$ K, from $r = 0.1\,R_{Sun}$ the bubble may need 10^{37}–10^{38} ergs to reach the surface, if we neglect the possibly favorable effect of involvement into a buoyant magnetic plasmoid here.

Figure 4 shows the rate of hydrogen deficiency of the bubbles relative to their local environments in dependence of the distance from the solar centre. Bubbles move upwards, therefore they represent the chemical composition of deeper regions, and

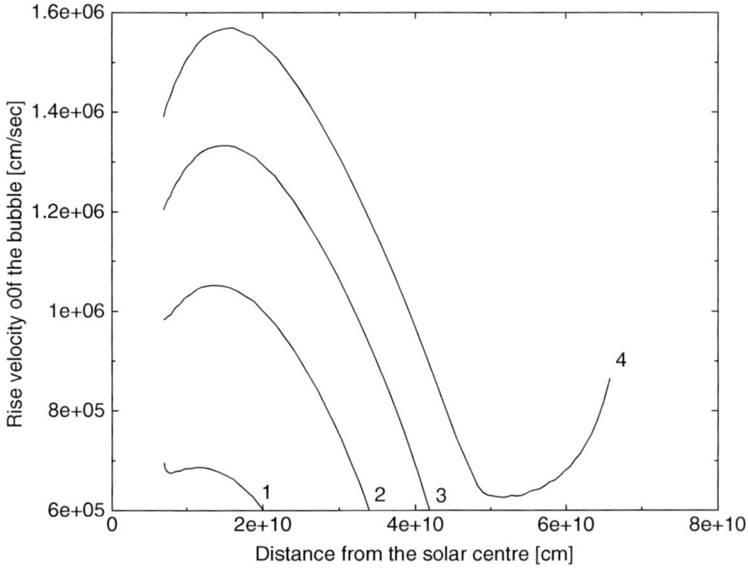

Fig. 3 The evolution of the velocity of the bubble with different initial sizes, for $T_0 = 9.0 \times 10^7$ K. Indices 1, 2, 3 and 4 refers to $R_0 = 10^6$ cm, $R_0 = 2 \times 10^6$ cm, $R_0 = 3 \times 10^6$ cm and $R_0 = 4 \times 10^6$ cm, respectively. The case with $\tau_{diff} = \kappa \rho_S^2 C_P R^2 / \left(16\sigma \left(T^3 - T_S^3\right)\right)$. The small waves around the end curve 4 are due a numerical effect, the linear interpolation applied in the solar model

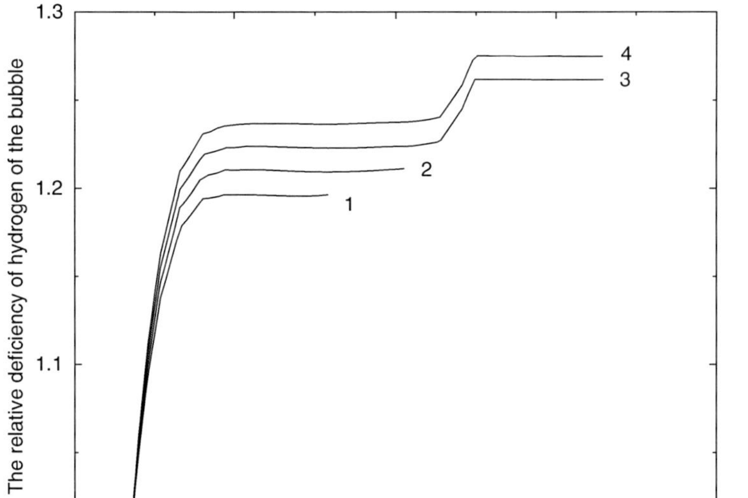

Fig. 4 The evolution of the relative rate of hydrogen deficiency of the bubble, for $T_0 = 9 \times 10^7$ K. Indices 1, 2, 3 and 4 refers to $R_0 = 10^6$ cm, 2×10^6 cm, 3×10^6 cm and 4×10^6 cm, respectively. The case with $\tau_{diff} = \kappa \rho_S^2 C_P R^2 / \left(16\sigma \left(T^3 - T_S^3\right)\right)$

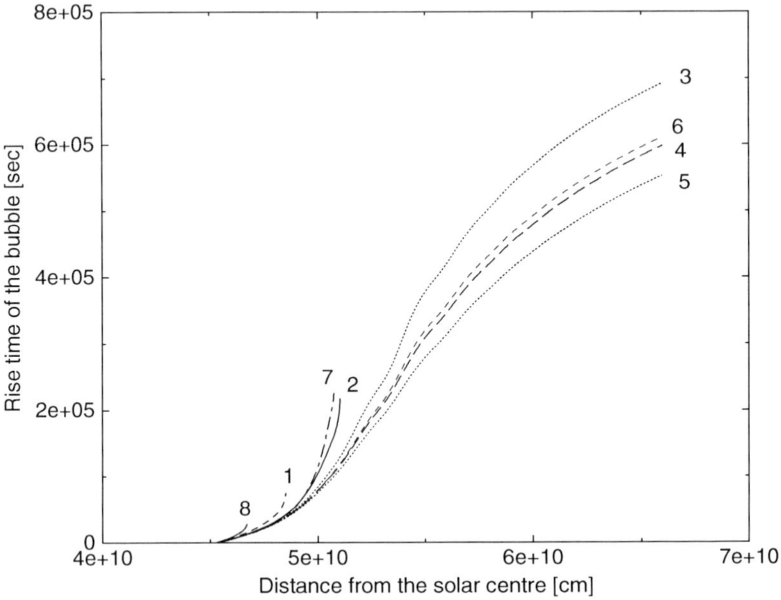

Fig. 5 The evolution of the relative temperature surplus of the bubble rising from $r_0 = 0.65R_{Sun}$ for $R_0 = 4 \times 10^5$ cm. Indices 1, 2, 3, 4, 5, 6, 7 and 8 refer to the cases with $T_0 = 5 \times 10^6$ K, 10^7 K, 1.5×10^7 K, 2×10^7 K, 3×10^7 K, 4×10^7 K, 5×10^7 K and 9×10^7 K, respectively. The case with $\tau_{diff} = \kappa\rho^2 C_P R^2 / \left(16\sigma \left(T^3 - T_S^3\right)\right)$

so their surfacing may be related to local chemical abundance anomalies. We will discuss the possible observational consequences of heavy element enhancements in solar flares shortly below.

Figure 5 shows the rise of the bubble for the case when the bubble starts to rise from $r = 0.1\,R_{Sun}$. One may notice that the increase of the initial temperature surplus of the bubble is not automatically helpful for the surfacing of the bubble. On the contrary, our calculations had shown that overly high heating causes faster cooling when the 'velocity' of radiative diffusion $v_{diff} = R/\tau_{diff}$ becomes higher than $v\,(= v_{rise})$, the bubble will cease to rise at lower distance from the solar centre. The stopping of the $R_0 = 4{*}10^5$ cm bubble for $T_0 = 5{*}10^7$ K will occur at $r_{final} = 5.07{*}10^{10}$ cm, and for $T_0 = 5{*}10^6$ K sooner, at $r_{final} = 4.85{*}10^{10}$ cm. This "cutting effect" of radiative diffusion may be effective in constraining the surfacing of the bubbles to a narrow range of heating, especially when the bubbles are formed not far below from the tachocline. Nevertheless, at deeper regions the cutting effect of radiative diffusion is less effective, but here other limiting factors may be effective, like heating energy input constraints. We note here that we obtained for the smallest energy surplus of the surfacing bubble a value of $\Delta Q_0 = 3.9{*}10^{31}$ ergs for $T_0 = 1.5{*}10^7$ K, when the bubble is formed and started its rise from $r_0 = 0.65\,R_{Sun}$. We note that we did not find bubbles reaching the convective zone for $R_0 > 4{*}10^5$ cm. This fact also means that $\Delta Q_0 > 3{*}10^{31}$ ergs is necessary for the bubbles to reach the subsurface regions, if we ignore here the possible role of being enveloped into a buoyant magnetic structure as a result of plasma interactions generating the bubble.

In Fig. 6 we plotted the rise of the bubbles for three cases with τ_{diff} proportional to ρ_S^2, starting from $0.1R_{Sun}$, from $0.4R_{Sun}$ and from $0.65R_{Sun}$. The common characteristics of these bubbles are that they are marginally able to reach the convective zone (and, therefore, the subphotospheric regions). Their path is close to each other near the bottom of the convective zone at $r = 4.9*10^{10}$ cm. The bubble rising from $0.1R_{Sun}$ have an initial energy surplus $\Delta Q_0 \sim 9.3*10^{37}$ ergs (dotted curve), the other one rising from $0.4R_{Sun}$ corresponds to $\Delta Q_0 \sim 4*10^{35}$ ergs (solid curve), the third one from $0.65R_{Sun}$ (dashed curve) to $\Delta Q_0 \sim 4*10^{31}$ ergs. The dashed curve runs to $\tau_{rise} \sim 6*10^5$ s at $r = 0.98R_{Sun}$. Bubbles surfacing from $0.65R_{Sun}$ have rising times $(5–6)*10^5$ s, almost independently from their initial temperature surplus.

We give some quantitative results regarding the arrival of hot bubbles into the subphotospheric regions in Table 4.

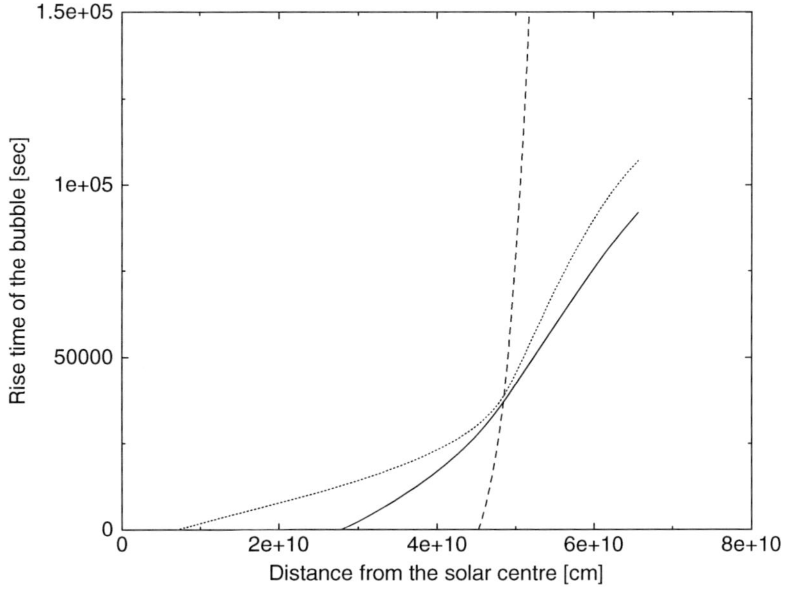

Fig. 6 The evolution of bubble rise from $r_0 = 0.1$, 0.4 and $0.65R_{Sun}$, shown in the dotted, solid and dashed lines, respectively. Dotted line: $T_0 = 9*10^7$ K, $R_0 = 5 \times 10^6$ cm. Solid line: $T_0 = 9*10^7$ K, $R_0 = 3 \times 10^6$ cm. Dashed line: $T_0 = 2*10^7$ K, $R_0 = 4 \times 10^6$ cm. $\tau_{diff} - \kappa\rho\rho S C_P R^2 / \left(16\sigma\left(T^3 - T_S^3\right)\right)$. The case of marginal surfacing

Table 4 The relations between subphotospheric bubble energies, bubble masses, final relative temperature surplus, final velocity and final radius

ΔQ_0[ergs]	ΔQ_{final}[ergs]	$m_{final}[g]$	T_{final}/T_S	v_{final}[km s^{-1}]	R_f[km]
$4.2 * 10^{31}$	$1.32 * 10^{30}$	$1.6 * 10^{18}$	1.01	0.6	35
$4.0 * 10^{35}$	$2.34 * 10^{33}$	$3.8 * 10^{20}$	1.10	3.9	209
$1.1 * 10^{38}$	$2.6 * 10^{34}$	$1.4 * 10^{22}$	1.04	4.3	680

We note that in Table 4 we presented only marginal cases, seeking the minimum initial energy surplus necessary to reach subphotospheric regions. Therefore, for larger initial energies, the hot bubbles may reach the near surface regions with higher speed and energy surplus.

Remarkably, the obtained energies, masses and sizes are in an apparently noticeable fit with the observed range of the same quantities characteristic to solar flares. This is the more interesting since the problem of flare energetics – as we will show here – has been never solved satisfactorily. It is many times emphasized that it is a belief that magnetic sources may supply the flare with enough energy. Priest and Forbes (2002, 317) express their belief that "The energy source [of solar flares] must be magnetic since all the other possible sources are completely inadequate". But Craig and McClymont (1999, 1045) stated that "the ability of any reconnection mechanism to explain the massive, explosive magnetic collapse of the solar flares remains questionable". Holman (2003) noted: It is generally believed that the flare energy is derived from the coronal magnetic field. However, we have not been able to establish the specific energy release mechanism(s) or relative partitioning of the released energy between heating, particle acceleration and mass motions. Machado (2001) wrote: "The most intriguing aspect of impulsive phase physics resides in the mechanism that leads to the release of, say, 10^{30} ergs in 10^2 s.The resolution of the impulsive phase enigmas will be addressed through a worldwide coordinated program of flare observations to start in the year 2000." Yet it seems that the worldwide coordinated program did not bring the long-awaited solution: "The energization process [of solar flares] is still unknown" (Masuda 2004). "Magnetic reconnection is generally believed to play a crucial role in solar coronal activity. A central paradox is that magnetic reconnection must occur at very small scales in order to be fast enough but must be directly affect the largest scales in order to matter" (Longcope 2005). This means that the resolution of impulsive phase enigmas still waits to be clarified.

The widespread conviction claiming that flare energies have to be supplied by magnetic reconnection is based on the notions that a) Magnetic energies may supply enough energy to flares, and, b) There is no any real alternative mechanism that may supply energy even comparably well. At first, we show that magnetic energies, in contrast to widespread beliefs, seem to be overly poor energy sources for solar flares. Secondly, we show that there is a powerful alternative energy source to energize flares.

When an engineer constructs the plan of a bridge, he must have plan the bridge to survive the possible largest challenges. Similarly, it is not the small, or moderate, but the large, and, first of all, largest flares that have a crucial importance regarding flare theories. Solar activity involves a wide range of phenomena on many spatial and temporal scales, corresponding to a wide variety of energy flows. In the case of a small flare, more energy source may be important, than in the case of largest flares, where the most effective energy source is indicated to be dominant. It is clear that we need to know about the most effective possible energy source of flares. Benz (2001) starts his review paper "Solar Flare Observations" with the introducing sentence: "Flares are caused by the release of magnetic energy up to some 10^{27} J (some 10^{34} ergs) in the solar atmosphere within a few minutes". This means that $E_f > 10^{34}$ ergs and $t_{imp} \sim 10^2$ $(<10^3)$ s. Li et al. (2005) found that the energy released by their white light flare is $\sim 10^{33}$ ergs – their flare is classified in X-rays as an X3-class flare. There are some dozens of flares of higher X-classes, one of the highest estimated is the 04/11/2003 being estimated as X28. Recently, it became clear that giant flares may be a feature of solar cycle, since giant flares are produced in both of the last solar activity cycles (Kane et al. 2005). Giant solar flares seem to have more than 10^{34} ergs already in the > 20 keV electrons, and, importantly, comparable or larger energies may be present in other forms (kinetic energies etc.). The frequency distribution of flares as a function of their X-ray energy are well defined over a broad energy range from 10^{27} to 10^{33} ergs (Kucera et al. 1997). These energy values do not contain the optical and mechanical energies that are many times indicated as larger, sometimes much larger. Nevertheless, as a cautious value, we can take that flare theories should explain the source of flare energy at least up to $E_f \sim 10^{33}$–10^{34} ergs, with $t_{imp} \sim 10^3$ s.

Now let us obtain a simple illuminating estimation of magnetic energies available for a big solar flare. With the magnetic field strength of an umbra $B_{umbra} \sim 3000$ G, its linear scale $L_{umbra} \sim 10^4$ km, its length $L_{length} \sim 10^5$ km, the magnetic energy $E_B \sim (B^2/8\pi) V_{tube}$ contained in a volume of a flux tube connecting a big sunspot pair $V_{tube} \sim 10^{27}$ cm^3 will be E_B (flux tube) $\sim 10^{34}$ ergs. Now it is a matter of fact that the fields usually appear to be fairly potential, so there isn't a big fraction of free energy. Many people think they cannot relax to a completely potential state in any case because of inductive time scales. Therefore, allowing a factor $\eta < 0.1$ for the fraction of magnetic free energy, one big sunspot pair, in case of a complete transformation of all its free energy content into the flare site within the short time of the flare, the available flare energy is still $E_f < \eta * E_B$ (flux tube) $\sim 10^{33}$ ergs. This simple estimation already shows that there is a problem with the magnetic explanation of flare energies supplied from the flux tube. Certainly, the total magnetic energy of a complete flux tube cannot be liberated in a flare also because observations do no indicate a characteristic disappearance of sunspots and their flux tubes after the onset of the flares. Now let us consider this point a bit more in detail.

Metcalf et al. (2005) had shown that there was an unusually large amount of free magnetic energy in NOAA AR 10486: E_B (available) $\sim (5.7 \pm 1.9) * 10^{33}$ ergs. This value involves the free magnetic energy of the whole active region. Moreover, as Kane et al. (2005) pointed out, even when the total available energy in the active region is comparable to the energy released during the flare, release of all that energy during the short duration of the flare is expected to affect substantially the magnetic field structure of the active region, and, apparently, there are no observations indicating large scale changes in an active region after a large flare. Therefore it seems that the active region does not offer enough magnetic energy to supply the largest flares. The magnetic field outside of active regions is so weak that it does not represent an effective energy source. Certainly, even the assumed inflows could not transport most of the flare energy from another active region. Therefore, if the magnetic field of a whole active region would prove to be insufficient to supply the flares, than models working with inflow transport of magnetic field would fail.

A closer look to the magnetic free energy content of active regions reveals a still more fundamental problem for the exclusively magnetic flare theories. The current sheets seem to be produced after the flare onset as a quickly elongated feature extending from the loop top. Sui and Holman (2003) found observational indications that a current sheet formed between the top of the flare loops and the coronal source moved outward at $\sim$300 km s^{-1}, showing an upward expansion of the current sheet during the early impulsive phase. Sui et al. (2005) noted that the large scale current sheet formed due to the fact that the coronal source moved outward at a speed of $\sim$300 km s^{-1}, while the loop top moved at only $\sim$10 km s^{-1}, thus, the current sheet must have continuously elongated. We find these observations as indicative that the current sheet is produced in the flare process itself. This observation seems to indicate it is the mechanism that produces the current sheet and the related phenomena that is the flare driver, and the current sheet is not the cause but an important consequence of the flare. This proposal seems to be underpinned by some other recent observations.

Schrijver et al. (2005) realized that in the active region 10486 it is the emergence of currents into preexisting active region field configurations that appears to be required to drive flaring. They noted that the free energy of active region fields available for flaring is not built up by persistent stressing of the surface field, but instead emerges with the field from below the photosphere. This result is underpinned by the recent results of Wheatland and Metcalf (2006) who were able to determine the free energy of the whole active region 10486 two hours before the flare as $(2.6 \pm 0.11) * 10^{33}$ ergs. Metcalf et al. (2005) determined that just after the impulsive phase the free energy of this active region is $(5.7 \pm 1.9) * 10^{33}$ ergs. They remarked that this suggests that the free energy was increasing prior or during the flare, but the errors in 2005 were overly large to allow such a statement to be verified. Nevertheless, their recent measurements decreased the errors and now their suggestion is underpinned with a significant probability. In the case if such results will be more definitely established, not only the mechanism producing the flare but the process producing $\sim 3 * 10^{33}$ ergs surplus in the free energy content of the active region magnetic fields within 2 h before and during the flare should be also explained. The same situation is found by Zhang (2001, Table 1): the magnetic flux of the flaring active region begins to increase more than an hour before the flare. Similarly, with the help of the THEMIS telescope and the Michelson Doppler Imager (MDI) on SOHO space probe, Meunier and Kosovichev (2003) presented observations showing that at the time of the flare, a sharp increase of the positive magnetic flux by 10^{21} Mx occurred during at least 1 h accompanied by strong flows both up and down especially at the time of the flare. Wang et al. (2004) also observed a rapid increase of magnetic flux in one polarity at the time of the flare, corresponding to a sudden emergence of new magnetic flux 10^{20} Mx hour^{-1} at the site of the flare. Wang (2005) pointed out that recent BBSO and MDI magnetograph observations demonstrate more and more evidences of rapid changes of photospheric magnetic fields associated with the core regions of flares and CMEs. Metcalf et al. 2005) refer to Uchida and Shibata (1988) as a reconnection scenario that predicted energy increase of the active region during the flare. Actually, Uchida and Shibata (1988) suggested that the supply of energy to the loop top comes from the chromosphere or transition region immediately before the flares in the form of relaxing fronts of magnetic twist of opposite signs traveling within the twisted flare loop itself. Therefore, their model does not explain the transport of energy into the active region from outside. In contrast, our proposal is able to explain both the production of this enormous free energy surplus and the flare with one simple factor: the fast emergence of flux tube from below the photosphere, in accordance with the findings of e.g., Schrijver et al. (2005).

Now let us underpin this proposal by a relevant argument and by an estimation. In the standard flare models it is a hypothetical 'catastrophic loss of equilibrium' that is responsible for the triggering of the flare (e.g., Priest and Forbes 2002). Recently, Lin (2005) suggested that the coronal accumulation of energy is built up by photospheric footpoint motions and the catastrophic loss of equilibrium is elicited by emerging flux tubes. Nevertheless, Schrijver et al. (2005) demonstrated that shear flows related to coronal free energy require appropriately complex and dynamic flux emergence within the preceding ~30 h and so they do not by themselves drive enhanced flaring. If photospheric footpoint motions are not able to support enough energy on the observed short period from the photosphere to the corona, another mechanism has to transport the photospheric energy into the corona, and this mechanism is directly related to the emerging flare loop itself. Now it is well known that flare loops are accelerated to speeds around 10 km s^{-1} for the period of flare onset (e.g., Bruzek 1964; Svestka 1968; Tsuneta 1993; Tsuneta 1997; Kundu et al. 2001). The speed of the flare tube ~10 km s^{-1} is remarkable especially in the light that the non-flaring emerging loops have much lower velocities. Fisher et al. (2000) reported on an average rise velocity of (non-flare) emerging flux tubes in the convective zone is around 0.01 km s^{-1}, corresponding to rise times through the convection zone 2–4 months. Zwaan (1992) noted that observations show loop tops passing through the photosphere with a speed of rise estimated at ~3 km s^{-1}. Schrijver et al. (1999) observed flux emergence at *chromospheric* heights with speeds averaged for a half an hour period as ~10 km s^{-1}. Caligari et al. (1995) calculated that at 13,000 km below the surface the radial velocity of the flux loop's summit is ~0.5 km s^{-1}. This theoretical result is confirmed by observations when Kosovichev et al. (2000) with the help of time-distance inversion methods determined that the speed of the emerging flux tubes in the *subphotospheric* regions are around 0.5–1.3 km s^{-1}. By numerical experiments of the emergence of magnetic flux, Archontis et al. (2004) determined that the main body of the rising flux tube acquires a rise speed of about 1.7 km s^{-1} while arriving at the *photosphere* yet still in the convective zone. By the analysis of observations, Spadaro et al. (2004) found that in the chromosphere the rising loop tops is characterized by velocities ~9 km s^{-1}. With the help of multi-height magnetic and velocity field measurements Choudhary et al. (2003) found that in the *photosphere* (as shown by Si I Dopplergram) the line-of-sight velocities characteristic of emerging flux is in the range 1.2 to –0.99 km s^{-1}, while in the chromosphere (as shown by He I Dopplergram) the velocities found are up to 5 km s^{-1}. Nagata et al. (2006) pointed out by high-resolution G-band observations that *photospheric* flux tubes move with a velocity 0.2–1.0 km s^{-1}. They pointed out that Stokes V profiles show flows with velocities up to $5\,\mathrm{km\,s^{-1}}$ inside and outside the flux tubes. It is also clear that the emerging flux tubes rise in an environment in which the gas pressure drops rapidly with height, therefore the loop expands favorably towards its top. The expansion of loops is highly pronounced in the chromosphere and the corona. Therefore, it is not easy to subtract the effect of expansion when someone aims to determine the rise speed of the main body of the rising flux tube. Yet we can summarize the above shown observational and theoretical results finding that they consistently show that the main body of an average, non-flaring flux tube ascends upwards in the subphotospheric regions with speeds $\sim$0.5–$\sim$2 km $\mathrm{s^{-1}}$.

The rise speed of the main body of the flaring loops around photospheric heights, in contrast, is much higher. Already Bruzek (1964, its Fig. 6) noted that there is a close association between loop prominences (flaring loops) and flares, and found that such flare-associated loop prominences rise with a speed $\sim$10 km $\mathrm{s^{-1}}$ from below the photosphere. Svestka (1968) realized that most of the observed slowly ascending limb flares may be explained by loop prominences ascending with speeds around 8 km $\mathrm{s^{-1}}$. The point is that most of the limb and disk flare loops are "slowly ascending" with very similar speeds, therefore, the emergence of flare loops with 8–10 km $\mathrm{s^{-1}}$ may explain most of the flares. Tsuneta et al. (1992) observed a limb flare with the Yohkoh X-ray telescope and found that the loop overlying above the flare loop starts to ascend at the flare onset with a speed of 10–30 km $\mathrm{s^{-1}}$, accompanied by a footpoint separation rate of a similar speed (a value larger by an order of magnitude of the footpoint separation speed of non-flaring emerging flux tubes). Tsuneta (1993) described the evolution of the flare in terms of flux tubes and their rise speeds. He found that the rapidly expanding flare loop of the 1991 December 2 flare appeared 5 minutes before the flare from below the photosphere, and its speed around flare onset is $\sim$10 km $\mathrm{s^{-1}}$, a value decreasing to 5 km $\mathrm{s^{-1}}$ after 30 minutes from flare maximum. Tsuneta (1997) in his Fig. 3 (upper panel) presented observations of the evolution of flare loop heights below and during the flare. These results tell that the flare loop had shown a quasi-constant rise speed from below the flare onset until $\sim$30 minutes after the flare maximum; from that time onwards its rise speed is decelerated. Kundu et al. (2001) observed that the main flaring loop of the 1993 November 11 event had emerged from below the photosphere just below the flare onset and started to rise with a speed about 8–10 km $\mathrm{s^{-1}}$. We think these observations demonstrate that flaring loops emerge from below the photosphere with a velocity 8–10 km $\mathrm{s^{-1}}$ that is much larger than the photospheric rise speed of quiescent, non-flaring emerging flux tubes.

It is important to observe that flaring loops seem to be accelerated not only by magnetic buoyancy, since non-flaring loops have similar fields but lower rise speeds. The existence of a non-magnetic acceleration factor seems to require the acceleration of material within and below the flux tubes. The average density in the subphotospheric regions is higher than $4{*}10^{-7}$ $\mathrm{gcm^{-3}}$. These flux tubes has a volume around $V{\sim}10^{27-28}\mathrm{cm^3}$. Now the acceleration of material extending to a similar volume to such high speeds requires significant energies $E_{acc} > 4{*}10^{32-33}$ ergs. This energy is in the same range required to fuel the magnetic free energy of flaring active regions by the recent results of Metcalf et al. (2005) and Wheatland

and Metcalf 2006). Their results, in agreement with many others Ishii et al. (1998, 2000, 2004; Kurokawa et al. 2002; Schrijver et al. 2005) involve that the free energy of the active region have to be supplied by emerging flux tubes. Now the most crucial emerging flux tube is just the flare loop. It is this context that offers a far-reaching context to our results that bubbles rise to the subphotosphere just by speeds $\sim$10 km s^{-1}, and their energies are in the range up to $\sim 10^{34}$ ergs and more. Now if exclusively magnetic flare theories cannot explain by a coronal 'catastrophic loss of equilibrium' the huge energies present in emerging flare loops and their acceleration to high speed for the time of flare onset, they fail to explain the most energetic aspect of flare phenomena.

There are many observations showing that flares are driven by emerging flux tubes. For example, Green et al. (2003) found that the majority of CMEs and flares occur during or after new flux emergence. Ishii et al. (1998, 2000, 2004) as well as Kurokawa et al. (2002) emphasized that "the emergence of twisted flux bundles is the energy source of strong flares". The same conjecture is drawn by Schrijver et al. (2005). Again, our argument works: if the flare is to be explained by an explosive coronal process liberating an enormous amount of radiative energy in a relatively short time and in a small volume; and if the energy source of the flare is the emergence of flux tubes from below the photosphere, than the only possibility is that the explosive coronal process is the result of the emerging flux tubes and not the other way around. Other observations underpin this argument. Seely et al. (1994) found that the energy is deposited in a small volume at the top of the flaring loop often as small (or smaller) than a single pixel (1,800 km$*$1,800 km). Zhang (2001) determined that the source of the Moreton wave in the X12/3B flare was originated between the photosphere and the upper chromosphere. Yamaguchi et al. (2003) found that the Moreton wave of the 1991 June 4 flare was emitted from a flare bright point that initially showed the form of a loop. We can identify the source of the Moreton wave with the looptop or the nearby region above it. Martin (2004) had shown that the X-class flares had a bright core in the chromosphere and the energy spread from this source. Movie 8 of Schrijver et al. (1999) shows the time evolution of the limb flare 1998 May 19. A slide show view of this movie demonstrates that the bright regions (associated with relatively dense and excited material) propagate consequently from below upwards before flare onset as well as during and after. Immediately before the flare onset, at 07:57:46, a bright vertical feature extends from the looptop of the flaring loop upwards. This bright vertical feature developed in relation to a feature ejected from the top of the flaring loop at 07:53:16 in the form of an ascending bright knot.

The above described observations show that the bulk energy of flares is supplied not from the free magnetic energy of the coronal region but in the form of currents and their hosting flux tubes emerging from below the photosphere. In the light of the above listed theoretical and observational results, we became aware of a fundamental problem: If the bulk energy of the flare is related to the emerging flux tubes, and if these flux tubes represent energy in the (sub)photospheric and chromospheric regions, how will be the energy of the flare localized into the looptop and the region above it? In other words: there is a missing link, a process concentrating the (sub)photospheric/chromospheric energy related to the flare loop into the coronal flare site. Fortunately, there are many important observations indicating the nature of this connection, like the ones we already mentioned above by Sui and Holman (2003) and Sui et al. (2005) indicating that the site of the primary energy release (the current sheet) develops from the looptop as an elongated vertical feature determined by the rise of the flare looptop with $\sim$10 km s^{-1} and the rise of the overlying loop with $\sim$300 km s^{-1}. These observations indicate that the site of the primary energy release of the flare (the current sheet) is enveloped into antiparallel field lines that are elongated with $\sim$300 km s^{-1}. In this context, the fundamental question of flare origin is: what is the mechanism driving the generation of this vertical magnetic feature?

If the energy of the flare is not supplied from the corona, one has to show which energy source can suffice, and provide also quantitative evidences. The detailed calculations that we realized considering the origin and development of the hot bubbles offer a new solution for this problem, nicely fitting to the findings of e.g., Schrijver et al. (2005). Table 4 shows that hot bubbles, when arrive to the subphotospheric regions, have tremendous energies, sufficing the energy requirement. The main obstacle is removed, since the basic problem of flare energy source can be solved within the frames offered by our model calculations, and so the main problem of the solar flares is resolved. Certainly, in order to make our promising result usable, a whole list of secondary problems should be solved. One of these accompanying problems is that we observe flares not in the subphotospheric, but in the coronal regions, and, apparently, with smaller masses. Let us present here some simple considerations that may relate our findings with more detailed observations.

The missing link is supplied by the observation that in the subphotospheric regions the hot bubbles are continuously accelerated in an environment in which the sound speed is smaller and smaller outwards. Our calculations indicate that hot bubbles may easily reach the threshold of sound speed where they may suffer sonic boom. If a hot bubble suffers a sonic boom, it will be destroyed by the developing 'compression wall' at the front of the generated shock wave (offering as a side result, a clue to the generation of flare related shock waves). Depending on local conditions, like the structure and strength of the magnetic fields, the material of the abruptly destroyed hot bubble compressed into the shock front will be transformed into a particle beam directed upwards, towards the loop top of the flux bundles that are carried by the hot bubble itself from below.

The point is that an energetic particle beam traveling between subphotospheric regions and loop tops with a velocity of 100–1,000 km s^{-1} is hardly visible until it interacts with the decelerating influence of relatively strong magnetic fields at the loop top. This is a crucial point in relating our numerical results to flare observations – and there are some promising answers offered. Already Kleczek (1964) pointed out, that energy in the form of corpuscular radiation is practically invisible in the photosphere and in the chromosphere. He referred to Warwick (1962) who argued that a 300 MeV particle beam may reach the photosphere from below a 300–500 km depth and we could not observe it in the optical region. Orrall (1964) commented this point, that "if the particles are injected into the loops at 1,000 km s^{-1}, we should not expect to see these in the corona – that is, in the coronagraph – against the sky. We should not expect to see them at all." Actually, the generation of particle beam injected towards the corona from below the photosphere should at first lift off the material of the photosphere. As a result, below the flare sites we will observe practically subphotospheric regions, i.e., somewhat hotter regions. Apparently, this is the phenomenon that is already observed by Machado and Linsky (1975) who remarked that "most of the flares show signs of photospheric heating during the flares. The energy of the photospheric heating is comparable to that of the chromospheric and coronal ones. This suggests that the energy source of solar flares is at least subphotospheric."

Energy considerations may turn to be fruitful when applied at the level of the photosphere and at coronal heights. Since we found that a particle beam may transport a large part of the energy content of the hot bubble from below the photosphere to the coronal loop top, the same amount of energy, $E_f \sim 10^{33}$–10^{34} ergs, is present in the photosphere as well as in the corona.

What is the size that is needed for a region with the temperature of photospheric regions, $T \sim 10^4$ K, to represent an energy content of e.g., $E_f \sim 10^{34}$ ergs? The answer of our model is simple. From the relation $E_f = cmT$, we obtain $m_b = E_f/cT \sim 5*10^{21}$g. With photospheric densities, $\rho \sim 4*10^{-7}$gcm^{-3}, we obtain for the linear size of the region $L \sim 10^9$cm, similar to the size of the flare kernels. On the other hand, since at the local sound speed the thermal and kinetic energies are equal, a region with similar size and having a velocity of the sound speed has a similar kinetic energy, $E_{kin} \sim 10^{34}$ ergs. Now we have a suitable mechanism to concentrate this energy into the shock front, compressing the hot bubble's material and generating the particle beam having much higher velocity: the sudden destruction of the high energy hot bubble, allegedly by the sonic boom (and/or by magnetic reconnection below the photosphere). Let us apply, as a first step, a most simple approach. Since, as we saw, a speed of $v \sim 10$km s^{-1} corresponds to $T \sim 10^4$ K, and the energy depends on the square of v, while depends linearly on T, 100 km s^{-1} for the particles corresponds to 10^6 K kinetic temperature at the flare site and 1,000 km s^{-1}–10^8 K. Actually, the site of primary energy release cannot be regarded as thermalized. Nevertheless, with some restrictions, one can speak of the temperature of the HXR source around the flare onset, and its value is found frequently in between 10^6 and 10^8 K; in energy units, 0.13–13 keV. Of course, nonthermal electrons corresponding to higher but unthermalized 'temperatures' represent a significant part of the energy of the flare. Therefore, on observational grounds we may require that the particle beam should correspond to a velocity $v_{particles} > 100$–1,000 km s^{-1}. By energy conservation we obtain that the mass carried by the particle beam $m_{particles}$ will be reduced in comparison to the mass of the subphotospheric hot bubble in a ratio.

$R_{mass} \sim \left(\left(10\,\text{km s}^{-1}\right)/v_{particles}\right)^2 < 10^{-2}$–$10^{-4}$, offering a mass for the particle beam $m_{particles} \sim m_b * R_{mass} < 5*10^{17}$–$5*10^{19}$g for the largest flares. Since the size of the primary energy release has a spatial scale $L \sim 10^4$ km, a mass of a large flare with $E_f \sim 10^{32}$ ergs will correspond to $m_{particles} \sim 5*10^{15}$g, a value that would increase the density at the loop tops to $n \sim 1.8*10^{11}$cm^{-3}. Again, this is consistent with observations. Already Zirin (1988, 409) noted that "it is really hard to understand how density can peak at the loop tops" in defiance to hydrostatic equilibrium. Actually, the density in the loop top may increase by an order of magnitude at flare onset in a duration less than a minute. Recently, Veronig et al. (2005) noted that plasma density is already enhanced at the flare onset, and increases for a peak density of n$\sim$ (1.3–2.2) $*10^{11}$cm^{-3}. In a looptop having a linear size $\sim 10^4$ km, such a density enhancement corresponds to a mass enhancement of $\sim 5*10^{14}$g.

We mention that recent theories and observations indicate the presence of subphotopsheric explosive processes in relation to solar eruptions (e.g., Hiremath, 2005b; Kosovichev and Duvall 2005). Some authors assumed that the subphotospheric events are related to subphotospheric magnetic reconnection. Not excluding this explanation, we propose also to consider that these events may be related to the dynamics of high-speed emerging flux tubes and the generation of particle beams. Kosovichev and Duvall (2003) already observed that the growth of active regions is characterized by multiple emergence of magnetic flux structures propagating *very rapidly* in the upper convection zone. In the context of our findings, the application of acoustic tomography to reveal the presence of high-speed subphotospheric bubbles before the flare beneath flare sites could serve with further observational evidences. Unfortunately, the time resolution of such measurements is at present very low, around 8 h (Kosovichev and Duvall 2004).

One of the remaining basic problems, as we indicated above, is how the X-type neutral point is formed as a result of the upward motion of the flare loop (Tsuneta 1997). In our model, it is natural to consider that it is the until now 'missing element', the energetic particle beam that elongates the field lines upwards, and so it generates the hard X-ray source at the looptop at the flare onset as well as the vertically elongating magnetic structure, in a way that it becomes elongated above the looptop afterwards due to particle beam injection from below. The expected result is an antiparallel, elongated magnetic semi-island growing upwards from the loop top, as a 'neck' with a 'head' which develops at the frontside into a separated 'plasmoid'. Now since the particle beam will not only push the field lines upwards, but at the same time a large part of the particle flux will follow the field lines bending towards the chromosphere, it will generate chromospheric evaporation, injecting also particle beams back into the loop top region from the chromospheric footpoints. These returning beams when transferring their momentum to field lines around the looptop will generate the elongated X-formed magnetic structures on the two sides ('arms'). We suggest particle beam injections from below as the mechanisms by which the observed characteristic X-shaped configuration as well as the antiparallel field line structure, i.e., the 'current sheet' develops (Fig. 7; from Fig. 5a of Tsuneta 1993).

The proposal presented here telling that particle beam injected from below into the top of the flare loop offers solutions to some problems, namely:

1. Our mechanism explains the basic problem how to generate the magnetic topology necessary for reconnection;
2. Our mechanism explains why the loop tops show upward elongated cusp-like structures instead of downward concave intrusions generated by particle beams injected from higher lying regions downward, as it is indicated by simulations (Forbes et al. 1989, Fig. 2; Forbes and Malherbe 1991, Fig. 5).

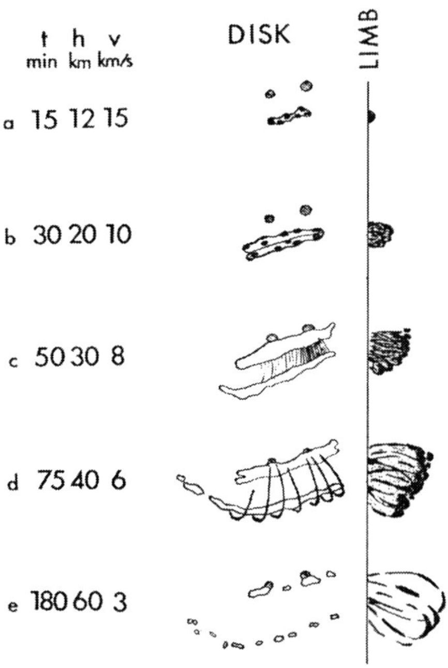

Fig. 7 The evolution of the magnetic field configuration typical at the onset of large flares at the disc and at the limb, after Bruzek (1964)

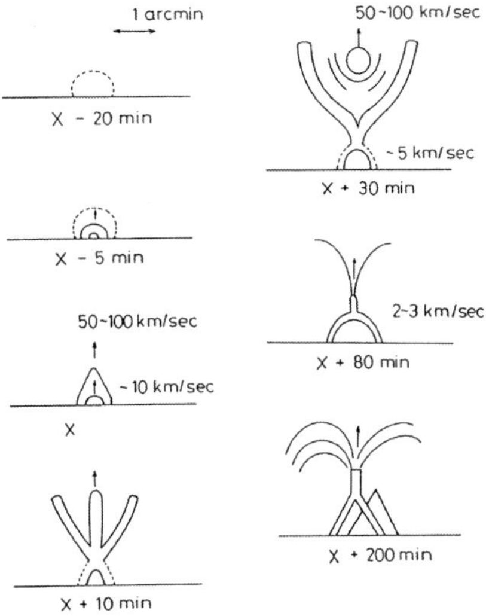

Fig. 8 The evolution of the magnetic field configuration around flare onset marked by x. From Tsuneta (1993)

Present-day pictures on the mechanisms generating the flares (e.g., Fig. 1 of Lin and Soon 2004) assume the presence of reconnection outflow injected into the loop top from above, but ignores the observable consequences that would be concave structures instead of the observed upward elongating antiparallel semi-islands and cusp-like structures;

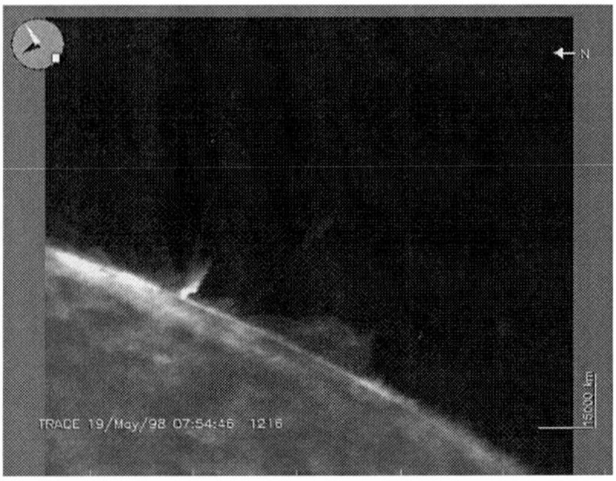

Fig. 9 Time evolution of a disc flare, at 19 May 1998, 07:54:46, after Schrijver et al. (1999), from their movie 8

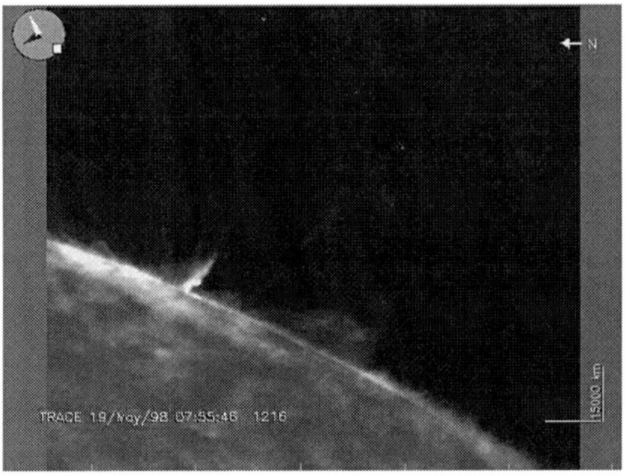

Fig. 10 Time evolution of a disc flare, at 07:55:46, after Schrijver et al. (1999), from their movie 8

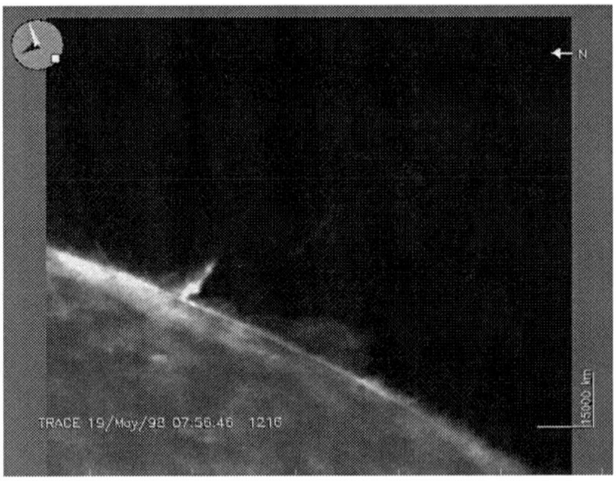

Fig. 11 Time evolution of a disc flare, at 07:56:46, around one minute before the flare, after Schrijver et al. (1999), from their movie 8. Our model interprets the upward moving bright know as the top of the particle beam travelling with 100–1,000 kms^{-1} injected from below

3. Our mechanism explains why the primary energy liberation occurs mostly above the loop tops but below the X-point. Saint-Hilaire and Benz (2002) had shown that most of the initial energy first appears as energetic electrons in the lower,

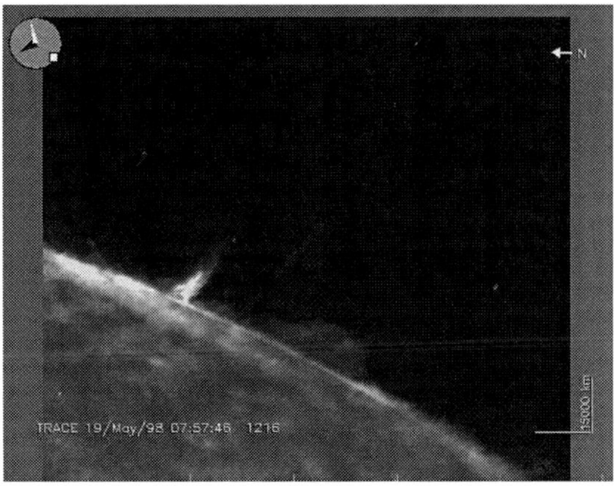

Fig. 12 Time evolution of a disc flare, at 07:57:46, around the flare onset. After Schrijver et al. (1999), from their movie 8

stationary part. Since the particle beams are injected in our model from below, the X-point is generated at the loop top and is driven upwards by the particle beam/plasmoid later on; therefore the primary energy release is produced in our model in between the loop-top and the X-point.

Our model has observational consequences also for the chemical composition of the flare material. Apparently, in the temperature range we explored here, nuclear heating does not help significantly the bubbles to reach the convective zone. As our results show, the reason is that the timescale of nuclear heating below T$\sim$1.5$*10^8$ K is larger than the timescale of cooling of the bubble, $\tau_{\mathrm{cool}} < \tau_{\mathrm{nucl}}$. Our tentative calculations indicated that thermonuclear runaway really might develop above T $>$ 1.5$*10^8$ K, where τ_{nucl} becomes lower than τ_{cool}. In that case, one could expect flare-related chemical anomalies with larger amplitude. Actually, the flare-related chemical anomalies represent a long-standing unsolved issue of solar physics (see e.g., Kerridge 1989; Sterling et al. 1993, Waljeski et al. 1994; Fludra and Schmelz 1999; Feldman et al. 2005). For example, Waljeski et al. (1994) presented results from measurements of soft X-rays (SXR) line and broadband intensities. They showed that for the observed active region the absolute abundances of the low first-ionization-potential (FIP) elements (Fe, Mg) are enhanced in the corona relative to the photosphere by a factor of 6–31, in a way that the abundances of the high FIP elements (e.g., Ne, O) are also enhanced by a factor larger than 1.75. Flare seed material plays a significant role in iron-rich gradual events. Recent measurements have shown that the heavy ion composition shows distinct differences from solar wind material, enhancements that are primarily due to the properties of the seed population of the flares (Mason et al. 2005). Although this question is still poorly understood, it seems possible to proceed by the development of new methods measuring the amount of absolute abundances of heavy elements in the flare. Our results predict that such absolute enhancements are actually present.

There are strong indications that the flare-related chemical anomalies represent not relative, but absolute enhancements (Grandpierre 1996). For example, the nitrogen enigma (Kerridge 1989) states that the $^{15}N/^{14}N$ rate is enhanced by 50%, from a value $2.9*10^{-3}$ of $3*10^9$ years ago to a present day value of $4.4*10^{-3}$ (Kerridge et al. 1991). It is just the opposite change of what the stellar evolution models predict. To produce the observed enhancement not only in the solar wind but also in the convective zone as a whole, would mean that the rate of this enhancement is so enormous that it exceeds the values by many orders of magnitude allowed by the standard models for the solar convective zone. This circumstance suggests that the solar surface is connected to the core by channels which are isolated from the convective zone, connecting the central regions with the subphotospheric regions directly. These central regions have to produce significant amount of heavy elements, like e.g., ^{15}N, which is possible only above 10^8–10^9 K, i.e., in a local explosive process. Our model can be tested independently when the chemical anomalies related to solar flares will allow determining absolute chemical abundances of the different elements present in the flare site. We add that recent solar models (Basu and Antia 2004; Bahcall and Pinsonneault 2004; Turck-Chieze et al. 2004; Guzik et al. 2005) pointed out that solar models evolved with standard opacities and diffusion treatment give poor agreement with helioseismic inferences for sound-speed and density profile, convection-zone helium abundance, and convection-zone depth. Varying the input parameters, none of the variations tried completely restores the good agreement attained using the earlier abundances. The problem is so severe that Guzik et al. (2005) recommended considering accretion of material depleted in the more volatile elements C, N, O, Ne and Ar. Our theory offers a natural alternative, again, without any further assumptions, by the occasional heavy element enhancements of some hot bubbles and related flare materials.

We mention that although the two crucial elements offered by our calculations for flare models, the high velocity hot bubbles in the subphotosphere and the particle beam injected from below the photosphere into the loop tops, are new propositions, they have the merit to draw together most of the results of competing flare models into one coherent picture. Namely, the newly arisen picture is consistent and complementary with the classical Hirayama (1974) model, with the Heyvaerts et al. (1977) emerging flux tube model, with the global picture of Priest (1995). It offers an energetically more suitable explanation for the generation of efficient acceleration of non-thermal particles (Hudson and Khan (1996) and is consistent with the current-centered flare model (Melrose 1997) with the idea that the particle beam interrupts the current at the loop top; it fits with the plasmoid model (Ohyama and Shibata 2002) since the plasmoid is the result of the interaction of the upward injected particle beam with the looptop. The model presented here suggests the presence of a primary nonthermal energy source in relation to preflare velocity fluctuations (Nigro et al. 2005) generated by the shock waves and particle beams since the sonic boom occurs before the flare and below the photosphere.

The results obtained here call attention to the principal possibility that local metainstability of generation of heated bubbles may explain the rigid rotation of some activity centers (Spence et al. 1993), as well as the existence of sunspot nests (Castenmiller et al. 1986; De Toma et al. 2000), hot spots (Bai et al. 1995), and active longitudes (Bai et al. 1995; Bai 2002, 2003). We note that the appearance of hot bubbles in the solar core may provide certain dynamism to the solar radiative interior and "very slow" mixing. The dynamic nature of the solar core (Grandpierre 1990, 1996, 1999) is indicated not only by the lithium problem (Deliyannis et al. 1998; Zahn 2001) and related problems with a need of a kind of mixing in the radiative interior, but also by the anomalously slow rotation of the core.

Now that we saw some successful explanation of a whole list of basic and yet unsolved problems, let us summarize here some predictions of our model that can be tested by future observations:

1. hot bubble(s) rising in the subphotosphere beneath the flare site accelerated to sonic speeds $\sim$8–10 km s^{-1};
2. shock waves are generated in the subphotosphere;
3. particle beam is generated in the subphotosphere and is injected upwards;
4. the material of the photosphere is lifted up before the particle beam, therefore the surface of the Sun is hotter underneath of the flare site than elsewhere;
5. flare material is enhanced in heavy elements not only as a result of selective electromagnetic processes;
6. reconnection topology is generated as a consequence of the process driving the coronal primary energy release of the flare, formed by particle beams injected from below;
7. measurement of the rise speed of the main body of flare loops;
8. the solar core have a dynamic nature that can be tested by e.g., detecting g-mode solar oscillations (Turck-Chieze et al. 2004).

Obtaining detailed data like local correlation tracking (LCT) based on speckle masking white-light images, near-infrared (NIR) continuum images at 1.56 mum, and G-band images could be helpful to test some of these points. Such data about flare-related photospheric flows are indicated to be a critical observational diagnostic for the evolution of magnetic fields in solar active regions (Deng et al. 2005).

Let us call attention to some new contexts that are shown as of interest in our understanding solar flares. For example, it is important to obtain more data on the time evolution of emerging flux tubes before and during flares (like Fig. 6 of Bruzek 1964 and Fig. 5a of Tsuneta 1993), also in dependence of depth/height beneath/above the photosphere. Movies of limb flares like that of Schrijver et al. (1999) could be especially helpful.

Our model is the first to solve some unsolved basic problems of solar activity. At the same time it indicates new problems, points out to new contexts, necessitates new observations and theoretical efforts, therefore it seems to be fruitful enough to consider it as worth to attention. In the face of the presented evidences, the model obtained here is indicated to be plausible. It answers some of the long-standing problems of solar physics and has predictions for observations that are planned to be realized in the near future. Revealing the presence of metastabilities in the solar core may help our understanding of the different types of instabilities, angular momentum dissipation, spin-down of the solar core and the dynamism arising from its plasma nature. The metainstabilities calculated in this paper may be directly relevant in our understanding of the generation of the solar, and, in general, stellar activity cycles.

Acknowledgements It is a pleasure to express our gratitude to Profs. Hudson, Priest, Forbes, Sturrock, Jakimiec, and Kálmán, for the useful notes that were helpful in the preparation of the final version. The author wishes to express his thanks to Dr. Hiremath for the valuable comments on the earlier version of the manuscript, and to his friend, Jean Drew for her galvanizing inspirations and lecturing the English.

References

Allen, C. W. 1963, *Astrophysical Quantities.* Third Edition. University of London, The Athlone Press, 161.
Alfven, H. and Arrhenius, G. 1976, SP-345 Evolution of the Solar System, Chapter 15, http://history.nasa.gov/SP-345/ch15.htm#250.
Alfvén, H. 1968, Second approach to cosmical electrodynamics, Ann. Geophysiq ue 24: 1.
Archontis, V., Moreno-Insertis, F., Galsgaard, K., Hood, A. and O'Shea, E. 2004, A&A 426, 1047.
Bahcall, J. N. and Pinsonneault, M. H. 2004, Phys. Rev. Lett. 92, 121301.
Bai, T., Hoeksema, J. T., and Scherrer, P. H. 1995, in *Helioseismology.* ESA SP (ed.), J.T. Hoeksema, V. Domingo, B. Fleck, and B. Battrick, p. 113.
Bai, T. A. 2002, AAS Meeting 200, No. 91.04.
Bai, T. 2003, ApJ 585, 1114.
Basu, S. and Antia, H. M. 2004, ApJ 605, L85.
Benevolenskaya, E. E. 1999, in: SOHO-9 Workshop *Helioseismic Diagnostics of Solar Convection and Activity*, Stanford, California, July 12–15, 1999.
Benz, A. O. 2001, Solar Flare Observations, in: *Encycl. of Astron. & Astrophys.*, Editor-in-Chief, Paul Murdin, Institute of Physics Publishing, Bristol and Philadelphia; Nature Publishing Group, London, New York and Tokyo, Vol. 3, 2529.
Bruzek, A. 1964, ApJ 140, 746.
Burgess, C., Dzhalilov, N. S., Maltoni, M., Rashba, T. I., Semikoz, V. B., Tortola, M. A., and Valle, J. W. F. 2003, ApJ 588, L65.
Caligari, P., Moreno-Insertis, F., and Schüssler, M. 1995, ApJ 441, 886.
Castenmiller, M. J., Zwaan, C., and van der Zalm, E. B. J. 1986, Sol. Phys. 105, 237.
Chang, T., Tam, S. W. Y., Wu, C.-C., and Consolini, G. 2003, Space Sci. Rev. 107, 425.
Chaplin, W. J., Elsworth, Y., Isaak, G. R., Miller, B. A., New, R., and Toutain, T. 2003, ApJ 582, L115.
Charbonneau, P. and MacGregor, B. 1992, ApJL 397, L63.
Choudhary, D. P., Suematsu, Y. and Ichimoto, K. 2003, Sol. Phys. 209, 349.
Craig, I. J. D. and McClymont, A. N. 1999, ApJ 510, 1045.
De Toma, G., White, O. R., and Harvey, K. L. 2000, ApJ 529, 1101.
Deliyannis, C. P., Boesgaard, A. M., Stephens, A., King, J. R., Vogt, S. S. and Keane, M. J. 1998, ApJ Letters 498, L147.
Deng, N., Xu, Y., Yang, G., Cao, W., Rimmele, T. R., Wang, H., and Denker, C. 2005, American Geophysical Union, Spring Meeting 2005, abstract #SP51C-05.
Ehrlich, R. 2007, J. Atmospheric and Solar-Terrestrial Physics, 69, 759–766.
Feldman, V., Landi, E., and Laming, J. M. 2005, ApJ 619, 1142.
Fisher, G. H., Fan, Y., Longcope, D. W., Linton, M. G., and Pevtsov, A. A. 2000, Sol. Phys. 192, 119–139.
Fludra, A. and Schmelz, J. T. 1999, A&A 348, 286.
Forbes, T. G., Malherbe, J. M. and Priest, E. R. 1989, Sol. Phys. 120, 285, Fig. 2.
Forbes, T. G., and Malherbe, J. M. 1991, Sol. Phys. 135, 361.
Friedland, A. and Gruzinov, A. 2004, ApJ 601, 570.

Garcia, R. A., Turck-Chieze, S., Couvidat, S., Ballot, J., and Mathur, S. 2004 in: Semaine de l'Astrophysique Francaise, meeting held in Paris, France, June 14–18, 2004, (eds.), F. Combes, D. Barret, T. Contini, F. Meynadier and L. Pagani, EdP-Sciences, Conference Series, meeting abstract.
Gervino, G., Lavagno, A., and Quarati, P, 2001, Nucl. Phys. A688, 126.
Goossens, M. and Tayler, R. J. 1980, Monthly Notices R. A. S. 193, 833.
Goossens, M. and Veugelen, P. 1978, Astron. Astrophys. 70, 277.
Goossens, M. 2003, *An Introduction to Plasma Astrophysics and Magnetohydrodynamics*, Kluwer, Dordrecht, 1, 15.
Gorbatsky, L. V. 1964, Astron. Zh. 41, 53 (in Russian; English translation in Soviet Astronomy-AJ, 1964, 8, 38).
Gough, D. O. and McIntyre, M. E. 1998, Nature 394, 755.
Green, L. M., Démoulin, P., Mandrini, C. H. and Van Driel-Gesztelyi, L. 2003, Sol. Phys. 215, 307.
Grandpierre, A. 1990, Sol. Phys. 128, 3.
Grandpierre, A. 1996, Astron. Astrophys. 308, 199.
Grandpierre, A. 1999, A&A 348, 993.
Grandpierre, A. and Ágoston, G. 2005, ApSS 298, 537.
Guzik, J. A., Watson, L. S. and Cox, A. N. 2005, ApJ 627, 1049–1056.
Haber, D. A., Toomre, J., Hill, F., and Gough, D. O. 1988, *Proc. Symp. Seismology of the Sun and Sun-like Stars*, Tenerife, Spain, 26–30 September 1988, ESA SP-286 (December 1988), 301.
Heyvaerts, J., Priest, E. R. and Rust, D. M. 1977, ApJ 216, 123.
Hirayama, T. 1974, Sol. Phys. 34, 323.
Hiremath, K. M. and Gokhale, M. H. 1995, ApJ 448, 437.
Hiremath, K. M. 2001, Bull. Astr. Soc. India 29, 169–181.
Hiremath, K. M. 2005a, personal communication.
Hiremath, K. M. 2005b, submitted to J. Astrophys. Astr.
Hiremath, K. M. 2009, Presented at the Fourth UN/ESA/NASA/JAXA Workshop on the International Heliophysical Year and Basic Space Science, Sozopol, Bulgaria, 2–6, June 2008.
Holman, G. D. 2003, ApJ 586, 605.
Hudson, H. S., and Khan, J. I. 1996, in: *Magnetic Reconnection in the Solar Atmosphere* (eds.), R. D. Bentley, and J. T. Mariska, ASP Conf. Ser. 111, 135.
Ishii, T. T., Kurokawa, H., and Takeuchi, T. T. 1998, ApJ 499, 898.
Ishii, T. T., Kurokawa, H. and Takeuchi, T. T. 2004, in: Stars as suns: activity, evolution and planets, Proc. of the 219th symposium of the I. A. U. (eds.), A. K. Dupree and A. O. Benz, San Francisco, CA: Astronomical Society of the Pacific (ASP), 2004, 693.
Ishii, T. T., Kurokawa, H. and Takeuchi, T. T. 2000, Publ. of the Astron. Soc. of Japan 52, 337.
Kane, S. R., McTiernan, J. M., and Hurley, K. 2005, A&A 433, 1133.
Kerridge, J. F. 1989, Science 265, 480.
Kerridge, J. F., Signer, P., Wieler, R., Becker, R. H., and Pepin, R. O. 1991, in: *The sun in time*. Tucson, AZ, University of Arizona Press, 389.
Kippenhahn, R. and Weigert, A. 1990, *Stellar Structure and Evolution*, Springer, 44.
Kleczek, J. 1964, in *AAS-NASA Symposium on the Physics of Solar Flares*, Washington 1963, NASA SP 50, 77.
Kosovichev, A. G., and Zharkova, V. V. 1998, Nature 393, 317.
Kosovichev, A. G. and Duvall, T. L. 2003, Amer. Geophys. Union, Fall Meeting 2003, abstract #SH22A-0186.
Kosovichev, A. G. and Duvall, T. L. 2004, American Astronomical Society Meeting 2004, #47.03.
Kosovichev, A. G. and Duvall, T. L. 2005, Amer. Geophys. Union, Spring Meeting 2005, abstract #SP51C-04.
Kosovichev, A. G., Duvall, T. L. and Scherrer, P. H. 2000, Sol. Phys. 192, 159.
Kucera, T. A., Dennis, B. R., Schwartz, R. A. and Shaw, D. 1997, 475, 338.
Kundu, M. R., Nindos, A., Vilmer, N., Klein, K.-L., Shibata, K. and Ohyama, M. 2001, ApJ 559, 443.

Kurokawa, H., Wang, T. and Ishii, T. T. 2002, ApJ 572, 598.
Li, J., Mickey, D. L. and And LaBonte, B. J. 2005, ApJ 620, 1092.
Lin, J. and Soon, W. 2004, New Astr. 9, 611.
Lin, J. 2005, American Geophysical Union, Fall Meeting 2005, abstract #SH11C-02.
Longcope, D. W. 2005, American Geophysical Union, Fall Meeting 2005, abstract #SH51D-05.
Machado, M. E. and Linsky, J. L. 1975, Sol. Phys. 42, 395.
Machado, M. E. 2001, Solar Flares: Impulsive Phase, in: *Encyclopedia of Astron. Astrophys.* Editor-in-Chief P. Murdin, Institute of Physics Publishing, Bristol and Philadelphia; Nature Publishing Group, London, New York and Tokyo, Vol. 3, 2537.
Martin, S. F. 2004, American Astronomical Society Meeting 2004, #47.11.
Mason, G., Desai, M., Mazur, J., and Dwyer, J. 2005, in 35th COSPAR Sci. Assembly, 1596M.
Masuda, S. 2004, in: *The Solar-B Mission and the Forefront of Solar Physics*. ASP Conf. Ser. 325 (eds.), T. Sakurai and T. Sekii, 341.
Melrose, D. B. 1997, ApJ 486, 521.
Metcalf, T. R., Leka, K. D., and Mickey, D. L. 2005, ApJ 623, L53.
Meunier, N. and Kosovichev, A. 2003, A&A 412, 541.
Nagata, S., Bellot R. L. R. and Katsukawa, Y. 2006, ApJ 638, 539.
Nigro, G., Malara, F., and Veltri, P. 2005, ApJ 629, L133.
Ohyama, M., and Shibata, K. 2002, in: *Multi-Wavelength Observations of Coronal Structure and dynamics – Yohkoh 10th Anniversary Meeting* (eds.), P. C. H. Martens, and D. Cauffman, Elsevier, 297.
Öpik, E. 1958, Irish Astron. J., 5, 36.
Orrall, F. Q. 1964, comment to the talk of Kleczek, in: *AAS-NASA Symposium on the Physics of Solar Flares*, Washington 1963, NASA SP 50, 79.
Paterno, L., Rapisarda, L., and Di Mauro, M. P. 1997, Astron. Astrophys. 322, 340.
Priest, E. R. 1995, in: *Flares and Flashes*, Proceedings of IAU Coll. 151, Sonneberg, Germany 1994 (eds.), Greiner, J., Duerbeck, H. W. and Gershberg, R. E., Lecture Notes in Physics 454, p. 3.
Priest, E. R. and Forbes, T. G. 2002, A&ARv 10, 313.
Ridpath, I. (ed.) 1997, *A Dictionary of Astronomy*, Oxford University Press, Oxford, New York, 450.
Rosenbluth, M. N. and Bahcall, J. N. 1973, ApJ, 184, 9.
Saint-Hilaire, P. and Benz, A. O. 2002, astro-ph/0210023.
Schrijver, C. J., Title, A. M., Berger, T. E., Fletcher, L., Hurlburt, N. E., Nightingale, R. W., Shine, R. A., Tarbell, T. D., Wolfson, J., Golub, L., Bookbinder, J. A., Deluca, E. E., McMullen, R. A., Warren, H. P., Kankelborg, C. C., Handy, B. N. and de Pontieu, B. 1999, Sol. Phys. 187, 261–302.
Schrijver, C. J., DeRosa, M. L., Title, A. M., and Metcalf, T. R. 2005, ApJ 628, 501.
Seely, J. F., Feldman, U., Doschek, G. A., Strong, K. T., Acton, L. W., Uchida, Y. and Tsuneta, S. 1994, In: Proceedings of Kofu Symposium, Kofu, Japan, 177.
Sewell, G. L. 1986, *Quantum Theory of Collective Phenomena*. Clarendon Press, Oxford, 4.
Spadaro, D., Billotta, S., Contarino, L., Romano, P. and Zuccarello, F. 2004, A&A 425, 309.
Spence, P., Walker, E. N., Halls, B., and Robertson, L. 1993, J. Br. Astron. Assoc. 103, 3.
Spruit, H. C. 2002, Astron. Astrophys. 381, 923.
Sterling, A. C., Doschek, G. A., and Feldman, U. 1993, ApJ 404, 394.
Sui, L. and Holman, G. D. 2003, ApJ 596, L251.
Sui, L., Holman, G. D., White, S. M., and Zhang, J. 2005, ApJ 633, 1175.
Svestka, S. 1968, in: Mass motions in solar flares and Related Phenomena. Proc. of the 9th Nobel Symposium (ed.), Y. Oehman, Stockholm: Almqvist &Wiksell, and New York: Wiley, 1968, 17.
Tayler, R. J. 1973, Monthly Notices R. A. S. 161, 365.
Toutain, T. and Kosovichev, A. G. 2001, in: *Proc. SOHO 10/GONG 2000 Workshop, 'Helio- and Asteroseismology at the Dawn of the Milleneum'*. Santa Cruz de Tenerife, Spain, 2–6 October 2000 (ESA SP-464, January 2001), 123.

Tsuneta, S., Hara, H., Shimizu, T., Acton, L. W., Strong, K. T., Hudson, H. S., and Ogawara, Y. 1992, Publs. Astron. Soc. Japan 44, L63.
Tsuneta, S. 1993, in: *The Magnetic and Velocity Fields of Solar Active Regions*, ASP Conference Series 46, 1993, (eds.), H. Zirin, G. Ai and H. Wang, 239
Tsuneta, S. 1997, ApJ 483, 507.
Tsytovich, V. N. 1970, *Nonlinear Effects in Plasma*. Plenum Press, New York London, 1.
Turck-Chieze, S. 2001, in: *Encyclopedia of Astronomy and Astrophysics*, Editor in-Chief Paul Murdin, Vol. 3, IOP Publ. Bristol, 2616.
Turck-Chieze, S., Garcia, R. A., Couvidat, S., Ulrich, R. G., Bertello, L., Varadi, F., Kosovichev, A. G., Gabriel, A. H., Berthomieu, G., Brun, A. S., Lopes, I., Pallé, P., Provost, J., Robillot, J. M., and Roca-Cortes, T. 2004, ApJ 604: 455–468.
Yamaguchi, K, Sakurai, T., Irie, M., Kumagai, K., Hagino, M., Miyashita, M., Shiomi, Y. and Hiei, E. 2003, Report of the National Astronomical Observatory of Japan, 6, 101.
Uchida, Y. and Shibata, K. 1988, Sol. Phys. 116, 291.
Veronig, A. M., Brown, J. C., and Bone, L. 2005, Adv. Sp. Res. 35, 1683.
Waljeski, K., Moses, D., Dere, K. P., Saba, J. L. R., Strong, K. T., Webb, D. F., and Zarro, D. M. 1994, ApJ 429, 909.
Wang, Q. L., Chen, Y. T., Cui, D. X., and Wang, W. P. 2004, in: *Intern. Union of Geodesy and Geophysics, Special Volume, Earthquake, Hazard, Risk, and Strong Ground Motion*, (eds.), Y. T. Chen, G. T. Panza, and Z. L. Wu, Seismological Press, 151, http://www.icce.ac.cn/shr_iaspei/docs/iugg_special_volume/iugg_p010_Wang.pdf.
Wang, H., Qiu, J., Jing, J., Spirock, T. J., Yurchyshyn, V., Abramenko, V., Ji, H. and Goode, P. R. 2004, ApJ 605, 931.
Wang, H. 2005, American Geophysical Union, Spring Meeting 2005, abstract #SH53B-01.
Wheatland, M. S. and Metcalf, T. R. 2006, ApJ 636, 1151.
Warwick, J. W. 1962, Publ. Astr. Soc. Pacific 74, 302.
Wolff, C. L. 2009, ApJ, 701, 686–697.
Zahn, J.-P. 2001, in: *Encyclopedia of Astronomy and Astrophysics*, Editor-in-Chief Paul Murdin, Vol. 3, IOP Publ. Bristol, 2563.
Zhang, H. 2001, A&A 372, 676.
Zirin, H. 1988, *Astrophysics of the Sun*, Cambridge University Press, Cambridge, 48, 402.
Zwaan, C. 1992, in: Sunspots: Theory and Observations,, (eds.), Thomas, J. H. and Weiss, N. O., Kluwer, 75.

How the Literature is Used A View Through Citation and Usage Statistics of the ADS*

Edwin A. Henneken, Guenther Eichhorn, Alberto Accomazzi, Michael J. Kurtz, Carolyn Grant, Donna Thompson, Elizabeth Bohlen, and Stephen S. Murray

Abstract The data holdings, usage and citation records of the NASA Astrophysics Data System (ADS) form a unique environment for bibliometric studies. Here we will highlight one such study. Using the citation and usage statistics from the NASA Astrophysics Data System, we study the impact of offering a paper as an electronic pre-print ("e-print") on the arXiv e-print repository, prior to its publication in a scholarly journal. We will address the following questions for astronomy: are people reading the e-prints from arXiv instead of the journal articles? Are e-prints read in a different way than journal articles? What is the impact of offering a paper as e-print prior to its publication in a scholarly journal? We will show that in astronomy, the e-prints are not being read instead of the journal article. As soon as the journal article is published, users prefer to read the article. Our analysis confirms that journal articles which were submitted as e-print on arXiv, prior to their publication, show higher citation rates than journal articles that were not submitted as e-print.

Keywords NASA/Smithsonian Astrophysics Data System · Citation statistics · Bibliometrics · e-prints

1 Introduction

We base our analysis on citation data in the ADS database, which are nearly 100% complete for the core astronomy journals. As part of the ADS, we create a mapping ("concordance") between the arXiv (see Ginsparg (2001)) e-print identifier and the identifier with which the published journal paper is known within the ADS

E.A. Henneken (✉), G. Eichhorn, A. Accomazzi, M.J. Kurtz, C. Grant, D. Thompson, E. Bohlen, and S.S. Murray
Harvard-Smithsonian Center for Astrophysics, 60 Garden Street, Cambridge, MA 02138, USA
e-mail: ehenneken@cfa.harvard.edu

*The ADS is funded by NASA Grant NNG06GG68G

H.J. Haubold and A.M. Mathai (eds.), *Proceedings of the Third UN/ESA/NASA Workshop on the International Heliophysical Year 2007 and Basic Space Science*, Astrophysics and Space Science Proceedings, DOI 10.1007/978-3-642-03325-4_12,

(the “bibcode”). This process of “bibcode matching” uses the metadata in the entries on arXiv (title, author names, submission information) to match the e-print with an existing ADS entry. Since here we look at e-prints that have been published, this concordance should ideally be 100% for our dataset. Unfortunately, titles and sometimes author lists have changed in the journal paper. From spot checks and from the 200 most cited papers we estimate that in reality the concordance achieved by the ADS is around 98% (see Kurtz et al. (2005)). This amount of incompleteness will have no influence on any of the outcome of our findings. We use citation data with a granularity of 1 month. Usage data are determined from “reads” by users through the ADS. The source of our data consists of the Astrophysics Data System usage logs and usage data provided by the arXiv team (going back to the beginning of 2000). For the ADS, we log all types of access by our users. An access “type” is related to which type of information viewed for an article. We define “reads” as the access events by users, where multiple information retrievals per log period for one article by one user is regarded as a single “read”. To rule out incidental use (e.g. by one-time users coming in via an external search engine, such as Google), we have taken the subset of users who query the database between 10 and 100 times per month. The ADS data was supplemented by usage made available to us by arXiv, which was similarly.

2 Results

2.1 Are People Reading the e-prints from arXiv Instead of the Journal Articles?

Figure 1 shows data based on the set of all articles from the 4 core journals in astronomy (Astrophysical Journal, Astronomical Journal, Monthly Notices of the Royal Astronomical Society and Astronomy & Astrophysics), from December 2004, for which the associated e-print was published in August 2004. We will refer to this set as *C4* (it consists of 118 papers). Figure 1 shows the reads per paper for the period of August 2004 through June 2006. The reads are from users with between 10 and 100 reads per month.

Figure 1 shows: (a) Reads of the arXiv e-print through the ADS very quickly drops to 0 after the publication of the journal article and (b) the half-life of the e-prints (the point where the use of an article drops to half the use of a newly published article) is considerably shorter than that of journal articles.

The ADS treats the e-prints and the journal articles equally, so the astronomical community is given the choice to read the e-print or the journal article. Figure 2 shows that the typical users prefer to read the journal article when this becomes available (Henneken et al. (2007)).

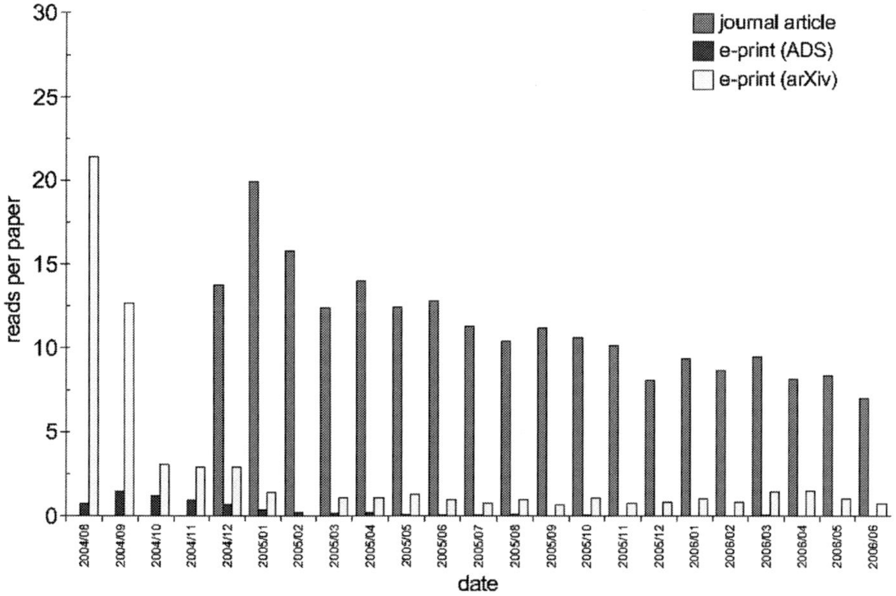

Fig. 1 C4 articles from December 2004, published 4 months after the arXiv e-print. Reads per paper from August 2004 through June 2006: reads through the ADS (article and e-print) and reads through arXiv

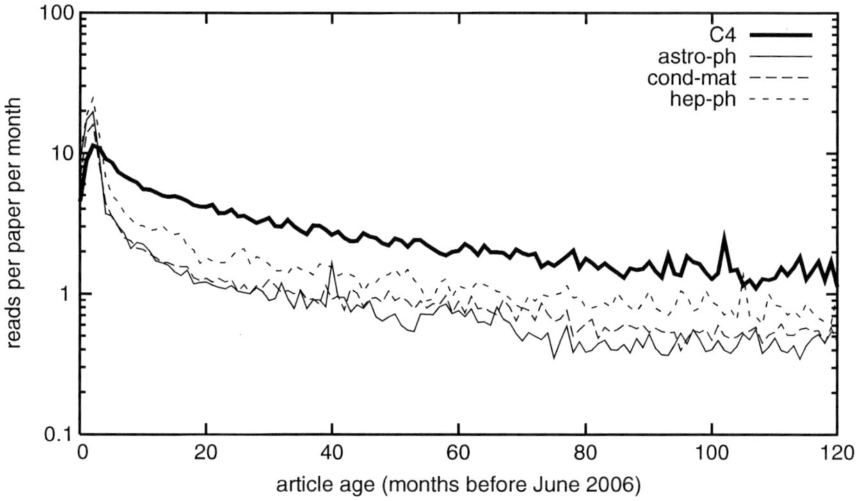

Fig. 2 The number of reads per paper per month: C4 through the ADS, astro-ph through arXiv

2.2 *Are e-prints Read in a Different Way than Journal Articles?*

To answer this question, we will look at the set of all e-prints in "astro-ph" and all articles from the *C4* journals. In order to be able to make a fair comparison, we only look at the articles that appeared as e-prints in the "astro-ph" category. So, each element in one set is linked to one and only one element in the other set. This set contains 38,171 papers.

Figure 2 shows the results for both sets. We have added the set of all papers published in the "cond-mat" and "hep-ph" categories as comparison.

The e-prints are heavily read just after their publication, but after that their readership quickly drops off to a fairly constant rate. On the other hand, the readership of journal articles stays high after their publication. The fact that the lines for journal article reads and for e-print reads parallel each other after about 12 months means that, from then on, the usage patterns evolve similarly, albeit with a factor of about five less e-print reads. The similarity between the arXiv and ADS readership after about 12 months shows that a significant amount of people use arXiv as their search engine.

2.3 *What is the Impact of Offering a Paper as e-print Prior to its Publication in a Scholarly Journal?*

The e-prints help journal articles to gain more visibility. This increased visibility is one of the reasons why journal articles that appeared as e-prints prior to their publication in the journal have higher citation rates than articles that did not (Henneken et al. (2006)).

The effect of e-prints on citation behavior is illustrated in Fig. 3. Here our dataset consists of articles published in the Astrophysical Journal and the Proceedings of the Astronomical Society of the Pacific, from the period of 1982–2002. Figure 3 shows the number of citations per paper, two years after the publication of the paper. The figure also shows the fraction of articles that was published as an e-print.

Figure 3 shows that with the introduction of the arXiv e-print repository in 1992, the number of citations per paper after two years starts to increase for ApJ. Although the number increases slightly for PASP as well, it is not as drastic as for ApJ. The reason behind this lies probably in the difference in subject matter and journal impact factor (or prestige). For example, there is more emphasis on instrumentation and catalogs in PASP.

Another way of looking at the impact of e-prints, is to select a period of time and determine the number of citations after publication as an ensemble average. We have taken all ApJ papers published in 1985–1987, and 1997–1999. We follow the citations to these papers for 5 years after publication of the paper. The period of 1985–1987 has been included as a comparison with the pre-arXiv era. We have determined how many citations an e-printed paper on average acquires as a function

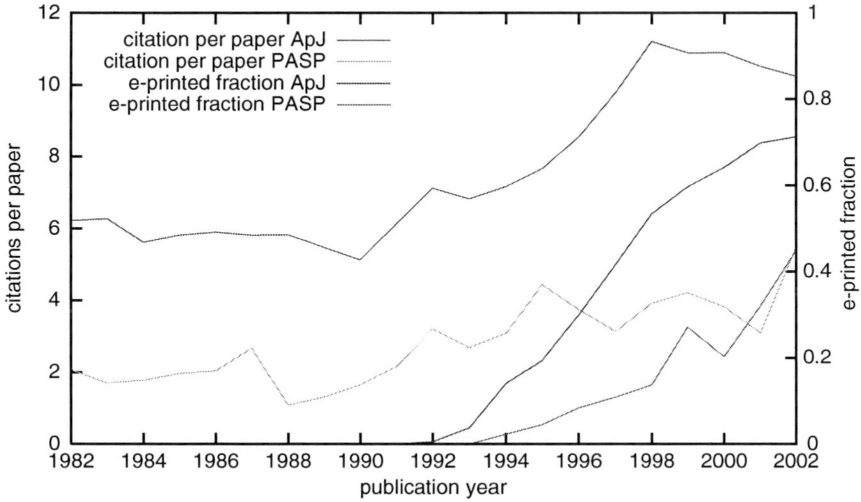

Fig. 3 Number of citations per paper two years after publication for ApJ and PASP

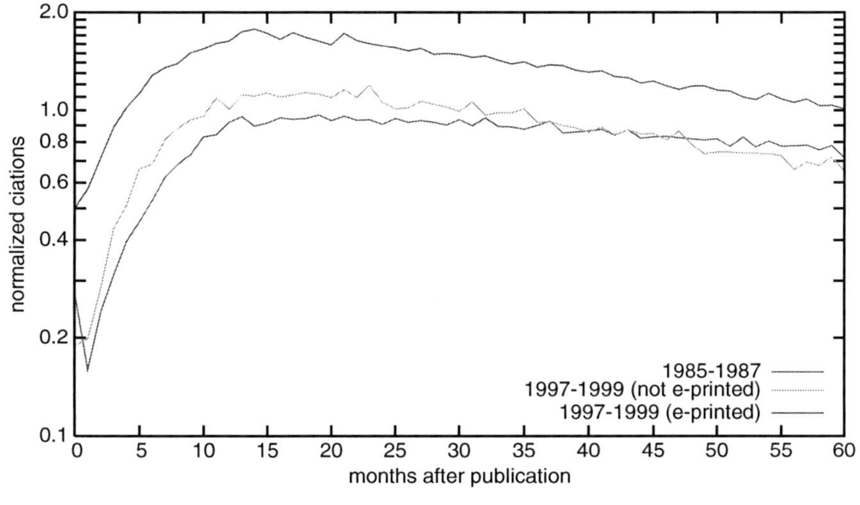

Fig. 4 Normalized citation count for ApJ papers

of time after publication, normalized with respect to the mean number of citations over the period of 5 years (over the entire dataset). We did the same for papers that were not e-printed. The result is shown in Fig. 4.

Figure 4 shows that ApJ papers from the period 1997–1999 which were not e-printed, follow the same citation trend as papers before the introduction of arXiv. Just like Figs. 3, 4 shows that the introduction of e-printing had a significant influ-

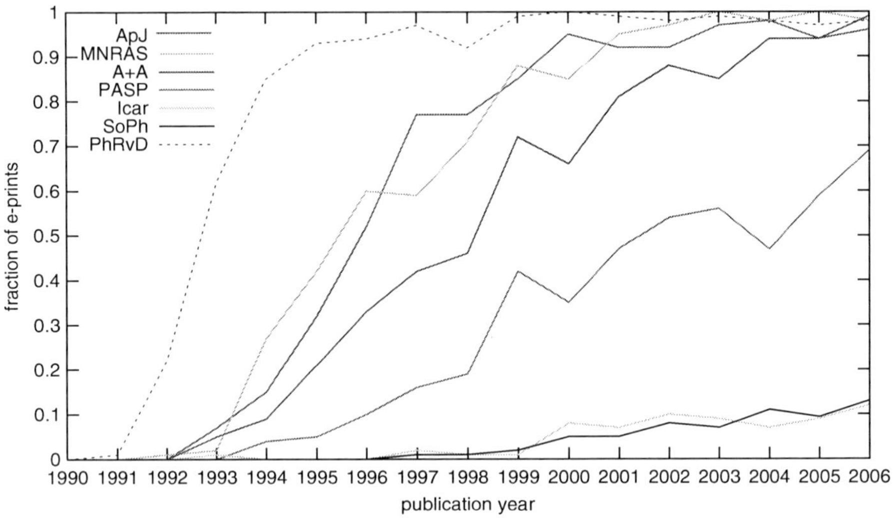

Fig. 5 Fraction of e-printed papers in the top 100 of most cited papers

ence on citation rates. The period of 1985–1987 was added as a comparison with the pre-arXiv era.

To illustrate the popularity of e-prints, we determined the fraction of e-printed papers in the top 100 of the most cited papers for a number of journals. We realize that within journals, one can also distinguish e-printing per discipline (see e.g. Schwarz et al. (2004)), but for our purposes we merely look at overall trends. The results are show in Fig. 5. Over 90% of the most cited papers in the main astronomy journals (the Astronomical Journal is not show here, but shows similar results) first appeared as e-print. This percentage is lower for the Publications of the Astronomical Society of the Pacific, probably because this journal contains many instrumentation papers, which represent a field that has not fully embraced the e-print culture. The fraction of e-prints is low for both Solar Physics and Icarus, representatives of planetary science research, which is another example of the dependency on discipline. The Physical Review D was included as an example of a journal where e-printing was very popular since the very beginning of arXiv.

3 Discussion

The members of the astronomical community consitute the lion's share of the journal article readers. For the core astronomy journals, we have shown (Figs. 1 and 2) that when given the choice, they prefer to read the published article over the e-print. We believe this is because the journal article has been refereed and is accepted as the "official" version. This is also a very important observation with respect to the economic component. The majority of astronomers has access to the online journals through institutional subscriptions, and therefore they have access to both e-prints and published articles. In other words, the e-prints have not undermined journal use in the astrophysical community! Quite the contrary, the e-prints help journal articles to gain more visibility. The increased visibility is one reason why e-printed articles receive higher citation rates. In Kurtz et al. (2005), it was shown that the higher citation rate of e-printed articles cannot be explained by assuming earlier availability as the sole reason. The quality of the papers is an additional motivation to offer it to the arXiv repository prior to publication in the journal ("self-selection bias"). This means that not only most papers are e-printed in the field of astronomy, but also that the best papers are e-printed. Over 90% of the top 100 of most cited articles in the main astronomy journals first appeared as e-prints.

References

Henneken, E.A., Kurtz, M.J., Eichhorn, G., Accomazzi, A., Grant, C.S., Thompson, D.,Bohlen, E., Murray, S.S., Ginsparg, P., and Warner, S., Learned Publishing, 20, 16 (2007)

Henneken, E.A., Kurtz, M.J., Eichhorn, G., Accomazzi, A., Grant, C., Thompson, D., Murray, S.S., Journal of Electronic Publishing, 9, 2 (2006)

Kurtz, M.J., Eichhorn, G., Accomazzi, A., Grant, C., Demleitner, M., Henneken, E., Murray, S.S., Information Processing and Management, 41(6), 1395

Kurtz, M.J., Eichhorn, G., Accomazzi, A., Grant, C.S., Demleitner, M., Murray, S.S., Martimbeau, N., and Elwell, B., Journal of the American Society for Information Science and Technology, 56, 111 (2005)

Schwarz, G.J., Kennicutt, R.C., Bulletin of the Astronomical Society, 36, 1654 (2004)

Ginsparg, P., in Electronic Publishing in Science II. Proceedings of Joint ICSU Press/UNESCO Conference, Paris, France. Available at: http://arxiv.org/blurb/pg01unesco.html (2001)

Photometric and Spectroscopic Studies of BW Eri

Desima Kristyowati, Hakim L. Malasan, and Hanindyo Kuncarayakti

Abstract New CCD photometric and spectroscopic studies of eclipsing binary BW Eridani are presented. BVRI photometric observations were carried out using Bosscha's 20-cm (f/10) GAO-ITB Remote Telescope System in 2006 and 28-cm (f/10) Schmidt–Cassegrain telescope in 2007. Low-resolution spectra ($R = 400 \sim 500$) were obtained using Bosscha's 45-cm (f/12) GOTO telescope equipped with Bosscha Compact Spectrograph (Malasan et al. 2001) in optical window. The investigation of B, V, R, I light curves by fitting method yields in temperature $7{,}480 \pm 2{,}950$ K and $5{,}200 \pm 875$ K, fractional radii 0.491 ± 0.126 and 0.280 ± 0.135, for the primary and secondary components, respectively. An inclination $89° \pm 2.2°$ is also deduced. We obtained the time for primary eclipse at HJD $= 2453769.1760 \pm 0.0118$ by Kwee–van Woerden method, which indicate period change. At the orbital phase of 0.955 and 0.511 the star's spectrum is consistent with spectral type G8V for the secondary and A7V for the primary, respectively.

Keywords Eclipsing binary · Photometry · Spectroscopy

1 Introduction

The eclipsing binary system BW Eri (SAO 169130; period 0.6384777 days) has been known as a candidate for a system caught in an out-of-contact state. The previous observations by Baade in 1976 and Duerbeck in 1977 were carried out with *UBV* photometers in European Southern Observatory, gave the evidence of period variation. The complete photometric study was first announced by Baade et al. (1982), who leads some conclusion i.e. BW Eri is a semi-detached system, with the primary filling its critical volume, component with very different surface temperatures, and a period variation that seems to indicate a mass transfer.

D. Kristyowati (✉), H.L. Malasan, and H. Kuncarayakti
Bosscha Observatory, Institut Teknologi Bandung, Lembang, Bandung 40139, Indonesia
e-mail: mydesima@gmail.com

H.J. Haubold and A.M. Mathai (eds.), *Proceedings of the Third UN/ESA/NASA Workshop on the International Heliophysical Year 2007 and Basic Space Science*, Astrophysics and Space Science Proceedings, DOI 10.1007/978-3-642-03325-4_13,

Baade et al. (1982) reported that the spectral type of the primary component from its colors is consistent with those of an unreddened A8V star; galactic latitude of BW Eri is $-46^\circ 98$. From the fractional luminosities, spectral type of secondary component yields G8, from the fractional radii yields G8V, and from mass ratio, K5V. No additional spectroscopic observations are available, so the spectral type that had deduced can be incorrect. Here, we present the results of photometric and first spectroscopic study of BW Eri.

2 Observations

2.1 Photometry

BVRI photometric observations were carried out in five nights (January–February 2006) using 20-cm (f/10) GAO-ITB Remote Telescope System (by DK, HK, HLM) equipped with an SBIG ST-7XE CCD. Additional observational data were collected in two nights (January 2007) using 28-cm (f/10) Schmidt–Cassegrain telescope (by DK, HK) equipped with an ST-8XME. Both observations were carried out at Bosscha Observatory. The data of the objects that were observed is listed in Table 1.

The comparison star and check star was selected from SAO Star Chart (1967), the period, colors, and spectral class from Baade et al. (1982). The ephemeris, Min I = HJD 2452664.5848, has been adopted from Dvorak (2004). The secondary minimum was well-sampled, but only half of the primary minimum has been observed due to weather instability and limited observable times.

2.2 Spectroscopy

Low-resolution spectra were obtained with 45-cm (f/12) GOTO Telescope equipped with Bosscha Compact Spectrograph (Malasan et al. 2001) and CCD ST-8XME in optical window at Bosscha Observatory. Five night's data cover essential orbital phase 0.955, 0.511, and 0.743.

Table 1 Variable, comparison, and check star data

Objects	Variable	Comp.	Check
SAO	169130	169134	169137
α_{2000}	$04^h\,06.61^m$	$04^h\,06.72^m$	$04^h\,07.02^m$
δ_{2000}	$-27^\circ 40.12'$	$-27^\circ 33.3'$	$-27^\circ 08.0'$
V	10.12−0.71	10.06	8.64
Period	0.6384777	–	–
B-V	+ 0.26 (phase 0.75)	+1.04	+1.26
Spec. class	A8+G8	K0	K2

The 300 mm^{-1} grating was used in the first order to obtain a spectrum covering $\lambda\lambda \approx$ 3,700–6,800Å. Exposure times for the object spectra were 600–1,800 s. FeNeAr (Hollow Cathode Tube) was used as comparison lamp.

3 Data Reduction

The photometric reduction was carried out with IRAF, Image Reduction and Analysis Facilities, *noao\digiphot\daophot* package included in (Stetson 1987). Standard aperture photometry reduction technique was applied.

With the use of IRAF, *twodspec* and *onedspec* packages, we applied the standard spectroscopy reduction.

4 Analysis

4.1 Light Curves

Synthetic light-curve fitting method has been applied to analyze the observed light curves. We fitted the light curves from observations with synthetic light-curves generated by Nightfall 1.42 (Wichmann 1999) and Binary Maker 3.0 softwares. Some preliminary parameters were taken from Baade et al. (1982). These parameters are mass ratio (q), inclination (i), fractional radii (r), and effective temperatures (T). Fitting process was implemented through trial and error by changing those parameters. An eye-fitting was done until we got the smallest χ^2 value of the fitting. The light curves and residuals as the results of fitting method are shown in Fig. 1.

Parameters that are suitable to these light curves represent the physical parameters of BW Eri. Table 2 shows the new parameters from this study.

The parameter q was fixed in order to make fitting converge rapidly. The gravity darkening exponent of the primary component which has a radiative envelope was assumed to be 0.25, and that of the secondary component which has a convective envelope was 0.08.

Limb darkening coefficients, $u(B)$, $u(V)$, $u(R)$ and $u(I)$ were adopted from Van Hamme (1993). The reflection effect was included. The albedos, $A_1 = 1.0$ for primary and $A_2 = 0.5$ for secondary was assumed. Stellar configuration at phase 0.75 is shown in Fig. 2.

To estimate the time of primary eclipse, we employed Kwee–van Woerden method to the V-band light curve (Fig. 3). This procedure yielded the time of primary eclipse at HJD $= 2453769.1760 \pm 0.0118$. This result indicates that the orbital period has changed; mass transfer was suspected to be the cause of this phenomenon.

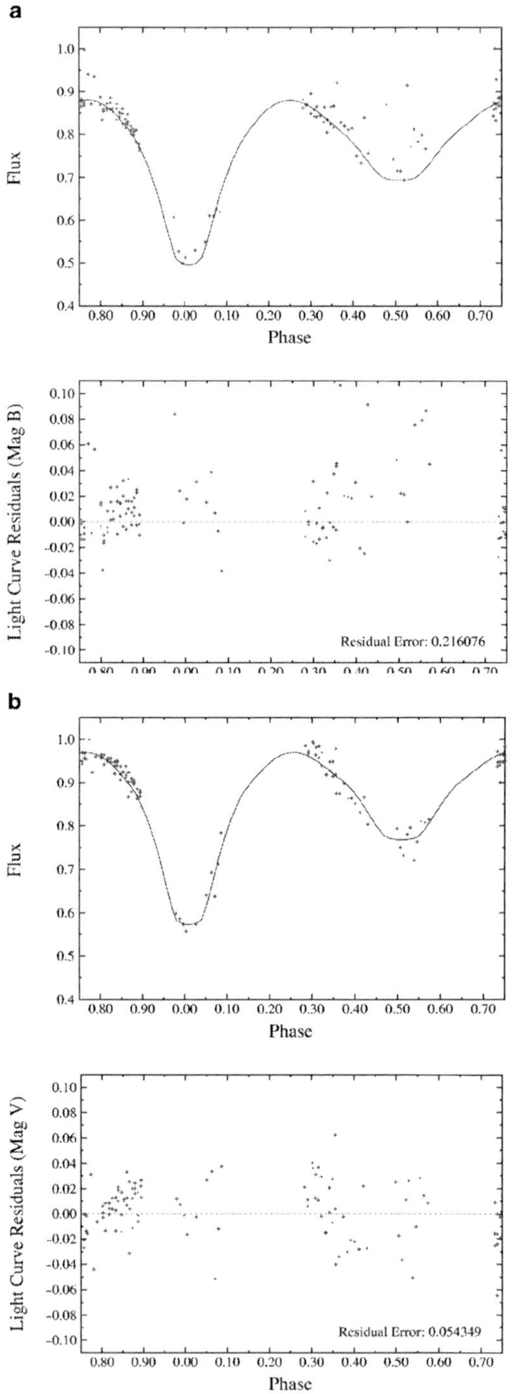

Fig. 1 Light curves fitting results and residuals in **(a)** B, **(b)** V, **(c)** R, and **(d)** I band

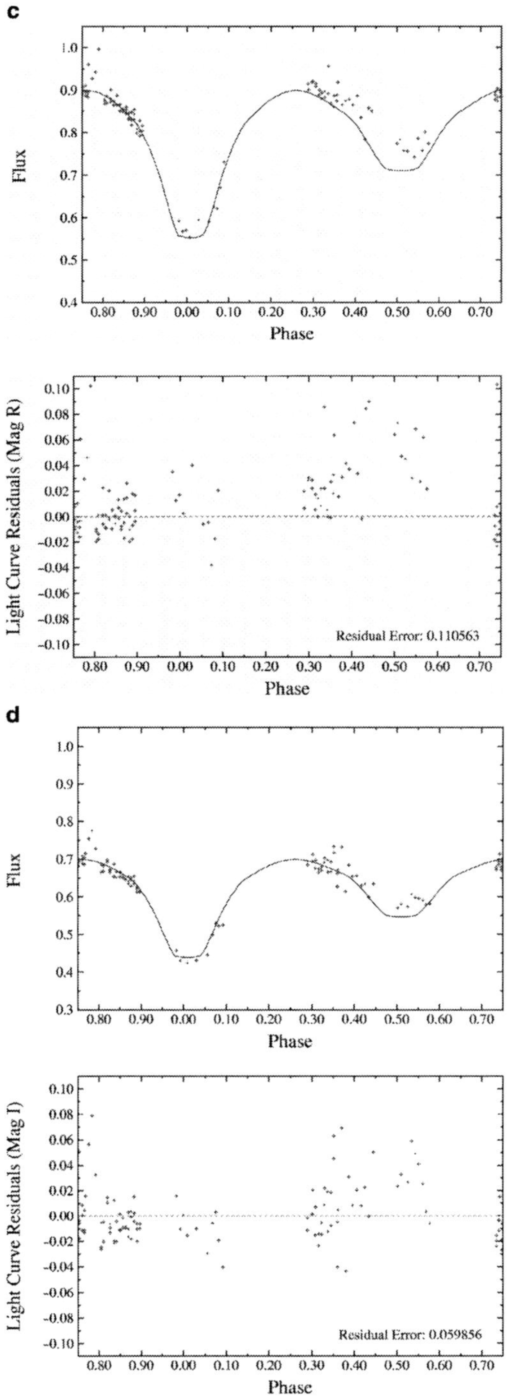

Fig. 1 (continued)

Table 2 Fitting parameters

Parameters	Primary	Secondary
i	89 ± 2.2	
β	0.25	0.08
$u(B)$	0.73	0.73
$u(V)$	0.58	0.68
$u(R)$	0.45	0.54
$u(I)$	0.38	0.49
T	$7{,}480 \pm 2{,}950$	$5{,}200 \pm 875$
A	1	0.5
Φ	2.619	2.720
q	0.36 ± 0.02	
$L(B)$	0.956 ± 0.012	0.044 ± 0.005
$L(V)$	0.937 ± 0.012	0.063 ± 0.005
$L(R)$	0.916 ± 0.012	0.083 ± 0.005
$L(I)$	0.898 ± 0.012	0.102 ± 0.005
r back	0.491 ± 0.126	0.280 ± 0.135
r side	0.466 ± 0.126	0.260 ± 0.135
r pole	0.437 ± 0.126	0.252 ± 0.135
r substellar	0.557 ± 0.126	0.296 ± 0.135

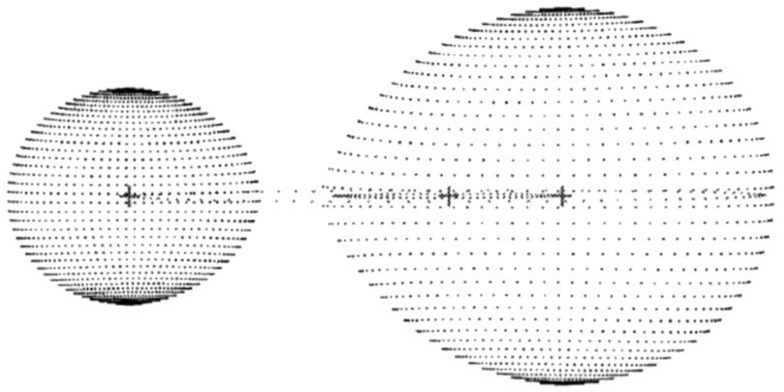

Fig. 2 Configuration of BW Eri system at phase 0.75

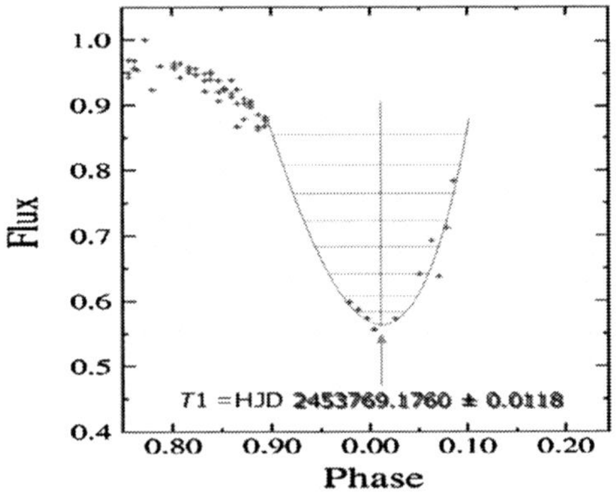

Fig. 3 Eclipse timing estimation

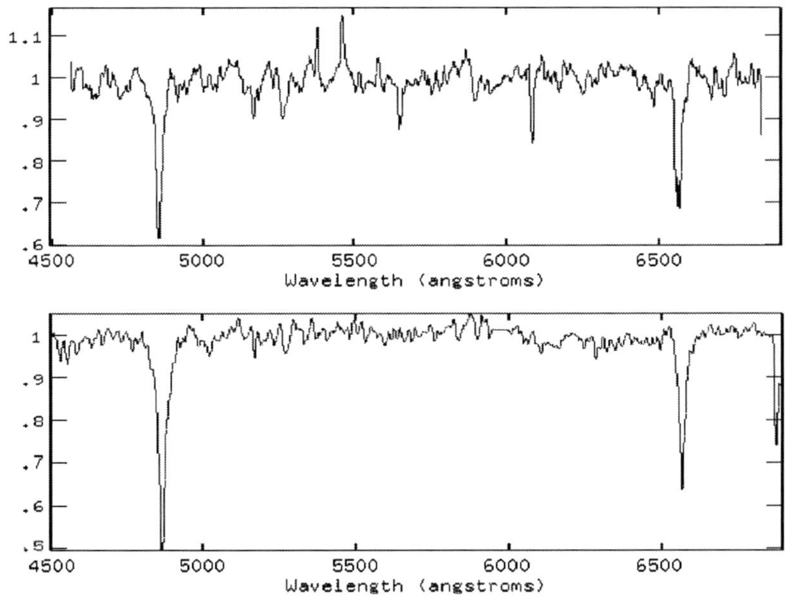

Fig. 4 The upper figure is the spectrum of BW Eri at phase 0.511, and the lower one is the A7V spectrum from Jacoby spectral atlas. Flux is normalized to unity

4.2 Spectra

The primary's spectrum was observed when the secondary was eclipsed by the primary; this is in accordance with orbital phase 0.511. The secondary spectrum was observed when the primary was eclipsed by the secondary; this is in accordance with orbital phase 0.955. With the use of Jacoby spectral atlas (Jacoby et al. 1984), we compare our spectra with the library to classify them into appropriate spectral classes. Spectral classification yields A7V for the primary and F8V for the secondary, respectively. The spectra are shown in the Figs. 4 and 5 below. But effective temperature yielded from light curve analysis indicated that the secondary seems consistent with G8 type.

5 Discussions

Analysis of the light curves using of fitting method to eclipsing binary data can be done to determine the physical parameters simultaneously. Complete coverage of light curves consisting of one primary minimum between two secondary minima. Nevertheless, it is felt that the primary minimum was shifted. The primary eclipse times was determined using the method of Kwee–van Woerden yield $HJD_I = 2453769.1760 \pm 0.0118$.

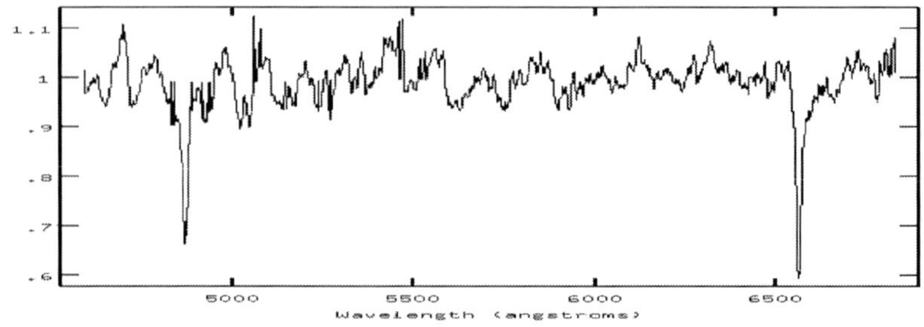

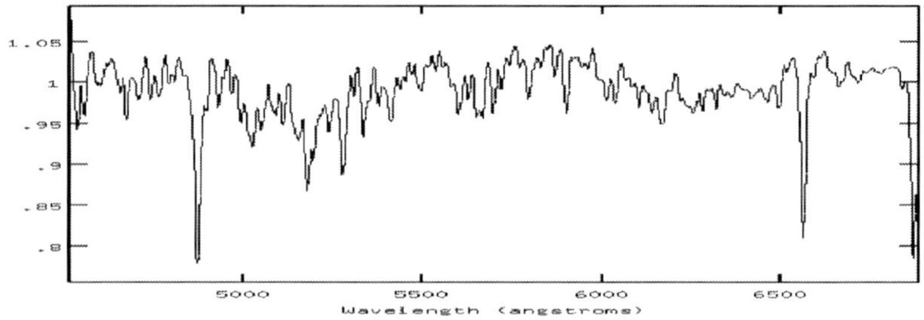

Fig. 5 The upper figure is the spectrum of BW Eri at phase 0.955, and the lower one is the F8V spectrum from Jacoby spectral atlas. Flux is normalized to unity

Because the mass ratio and inclination were kept fixed in the analysis, there is a significant change in fractional luminosities of secondary from the preliminary ones. Previous study (Baade et al. 1982) revealed that BW Eri was a semi-detached eclipsing binary with the primary component filling its critical lobe, and there was mass transfer indication to the secondary. The changes of fractional luminosities and radii serve as the evidences of mass transfer and period changes.

There is inconsistency of secondary spectral type obtained from spectrum and light curve. This is expected that at orbital phase 0.955.

Acknowledgements We would like to thank Professor M. Kitamura for constant encouragement. The authors also thank Dr. C. Kunjaya and Dr. M. Putra for fruitful discussions, and Dr. D. Herdiwijaya for generously allocating observation time at the Bosscha Observatory.

References

Baade, D., et al.: A photometric study of the short-period eclipsing binary BW Eridani. Astrophys. & Space Science. 88, 197 (1982)

Dvorak, S.W.: Times of minima for neglected eclipsing binaries in 2003. Inf. Bull. Var. Stars. 5502 (2004)

Jacoby, G.H., et al.: A library of stellar spectra. Astrophys. J. Suppl. Ser. 56, 257–281 (1984)

Malasan, H.L., et al.: The Bosscha Compact Spectrograph. In: Hashimoto, O. (ed.) Proceedings of Workshop on Astrophysics with High-resolution Spectrograph in Japan, Japan, 2001

Stetson, P.B.: DAOPHOT: a computer program for crowded field stellar photometry. Publ. Astron. Soc. Pac. 99, 191–222 (1987)
Van Hamme, W.: New limb darkening coefficients for modeling binary star light curves. Astron. J. 106, 2096–2117 (1993)
Wichmann, R. In: http://www.lsw.uniheidelberg. de/_rwichman/Nightfall.html/ (1999)

Near Infrared Excess Energy in Binary System V367 Cygni

Saraj Gunasekera

Abstract Spectral energy distribution of the Serpentid type binary V367 Cyg was obtained using several previous photometric measurements made on this system in different spectral bands. We found Near IR excess starting from 3 μm and this excess flux is attributed to the free–free emission from the mass accretion disk of the binary system. We adopted the temperature of primary component as 8,000 K. We added the free–free emission flux of the circumstellar disk to the black body energy of the primary component to find a best fit for the observed near infrared excess flux. In this fitting we left the electron density of the circumstellar disk n_e of the free–free emission as a free parameter. We found that volume emission measure of the circumstellar disk is $\sim 9 \times 10^{59}$ cm^{-3}.

Keywords Serpentid stars · V367 Cyg · Circumstellar disk · Free–free emission

1 Introduction

V367 Cygni, BD + 38°4235, V = 7.04^{m}, B = 7.66^{m}, P = 18.5972 days, Sp = A7Iapevar, RA = 20^{h}47^{m}59^{s}.6, DEC = +39°17′15″.7 (J 2000) is a bright eclipsing binary system which is included in the group of Serpentids by M J Plavec in 1980. W Serpentis star is a close binary star system where matter is being transferred very rapidly from one star to the other. It is believed that V367 Cygni is in a phase of rapid mass transfer. Its optical spectrum resembles to an evolved A-type star. Only the spectrum of primary component is observable and it exhibits a shell-like spectrum, an indication of a gaseous stream of material surrounding the system. The underlying stellar spectrum is dominated by sharp shell lines mostly of singly-ionized shell lines. MgII λ4481 is the only strong and clearly visible stellar line in the optical spectrum. The primary component which is more massive component at the outset of the mass losing to the secondary component has now become

S. Gunasekera (✉)
Arthur C. Clarke Institute for Modern Technologies, Katubedda, Moratuwa, Sri Lanka
e-mail: saraj@accimt.lk; sarajgunasekera@yahoo.com

H.J. Haubold and A.M. Mathai (eds.), *Proceedings of the Third UN/ESA/NASA Workshop on the International Heliophysical Year 2007 and Basic Space Science*, Astrophysics and Space Science Proceedings, DOI 10.1007/978-3-642-03325-4_14,

less massive component. The mass gainer, secondary component, has become more massive and hotter but like in other W Serpentis stars it is embedded in a geometrically and optically thick accretion disk which almost entirely obscure the secondary from viewing.

Pavlovski et al. (1992) developed a code in which the influence of an optically and geometrically thick disk, surrounding the mass accreting component has been taken into account. Based on this code they arrived at a solution to the light curve of the system with mass ratio q = 0.56 which corresponds to the accretion disk radius of 0.43A.

Zola and Ogloza (2001) using the Wilson–Devinney code also modeled the light curve with and without accretion disk. They found a contact configuration for the no-disk model and a semidetached model for the disk model which explains most of the observed features of V367 Cygni. From their light curve model they found the radius of the accretion disk $R_d = 23R_\odot$.

Taranova has also found the near IR excess emission at 3.5 and 5 μm and it was attributed to the emission from disk-shaped ionized shell.

Having gone through the previous studies done on this system and obtained the photometric magnitudes of different wavelength bands provided by several observations we obtained the spectral energy distribution of the system. In this study we also adopted absolute parameters of V367 Cygni found in several research papers for our calculations.

2 Analysis of Photometry Data

In the first part of the data analysis the Spectral Energy Distribution (SED) of the V367 Cygni is presented. The photometry data obtained from several previous studies are used to derive the SED of the binary system and given in the Table 2.

The magnitudes of the V367 Cygni obtained from previous observations are converted in to the observed flux using the flux of the zero magnitude star given in the Table 2.

Observed flux f_λ can be calculated using the magnitude definition of a star as given below.

$$m_\lambda - m_{\lambda(zero)} = 2.5 \log \frac{f_{\lambda(zero)}}{f_\lambda}$$

$$\log f_\lambda = \log f_{\lambda(zero)} - \frac{m_\lambda}{2.5} \quad \text{since } m_{\lambda(zero)} = 0$$

Interstellar extinction corrected magnitude $m_{0\lambda}$ is defined as,

$$m_{0\lambda} = m_\lambda - A_\lambda \text{ where } A_\lambda \text{ is the extinction}$$

$$m_{0\lambda} - m_{\lambda(zero)} = 2.5 \log \frac{f_{\lambda(zero)}}{f_{0\lambda}}$$

$$\log f_{0\lambda} = \log f_{\lambda(zero)} - \frac{m_\lambda - A_\lambda}{2.5} \quad \rightarrow \qquad (1)$$

Extinction in the visual band is found to be proportional to the color excess E_{B-V}.

$$A_v = RE_{B-V} \quad \text{where R} = 3.1$$

Analyzing the low resolution spectra provided by IUE (International Ultraviolet Explorer satellite) Hack et al. have found that the system is heavily reddened with E_{B-V} between 0.5 and 0.55.

In this study we adopt $E_{B-V} = 0.5$. Therefore $A_V = 3.1 \times 0.5 = 1.55$

A_λ for each wavelength band is calculated using the values A_λ/A_V given in the Table 1. Using (1) interstellar extinction corrected flux for observed magnitude of the each band is calculated and is given in the Table 2.

Spectral Energy Distribution (SED) of the system is depicted in the Fig. 1. From the absolute parameters of V367 Cygni derived from the light curve modeling within the disk model by Zola and Ogloza (2001) and Pavlovski et al. (1992) we adopted the effective temperature of the primary component to be 8,000 K. A black body curve with the same temperature was generated using a computer program and

Table 1 The near infra red extinction law

A_V/E_{B-V}	3.10	A_V/A_V	1.000
E_{B-V}	1.00	A_B/A_V	1.323
E_{V-J}	2.28	A_J/A_V	0.265
E_{V-H}	2.62	A_H/A_V	0.155
E_{V-K}	2.82	A_K/A_V	0.090
E_{V-L}	2.96	A_L/A_V	0.045
E_{V-M}	3.02	A_M/A_V	0.026

Table 2 Flux calibration of V367 Cygni

Band	λ_0 (μm)	Mag.	Flux (zero mag) (W cm$^{-2}\mu$m^{-1})	Extinction (A_λ/A_v)	(A_λ)	Obs. flux$_{\text{ext corrt.}}$ (W cm$^{-2}\mu$m^{-1})	Log obs. flux$_{\text{ext corrt.}}$ (W cm$^{-2}\mu$m^{-1})
B	0.440	7.66[a]	7.20E-12	1.3230	2.0507	4.11E-14	−13.386408
V	0.550	7.04[a]	3.92E-12	1.0000	1.5500	2.50E-14	−13.602714
R	0.640	6.65[b]	1.76E-12	0.8000	1.2400	1.21E-14	−13.918487
I	0.900	6.34[b]	8.30E-13	0.5900	0.9145	5.61E-15	−14.251122
J	1.250	5.20[b]	3.40E-13	0.2650	0.4108	4.13E-15	−14.384221
H	1.650	5.00[b]	1.28E-13	0.1550	0.2403	1.60E-15	−14.796690
K	2.200	4.80[b]	3.90E-14	0.0900	0.1395	5.33E-16	−15.273135
L	3.400	4.51[c]	8.10E-15	0.0450	0.0698	1.36E-16	−15.867615
M	5.000	4.60[c]	2.20E-15	0.0260	0.0403	3.30E-17	−16.481457
F_8	8.000					6.14E-18[d]	−17.211832
F_{12}	12.000					2.35E-18[d]	−17.629487
F_{25}	25.000					2.76E-19[d]	−18.559878

[a]S. Zola and W. Ogloza
[b]Mass, USNO
[c]O.G. Taranova
[d]IRAS (Infrared Astronomical Satellite) Measurements

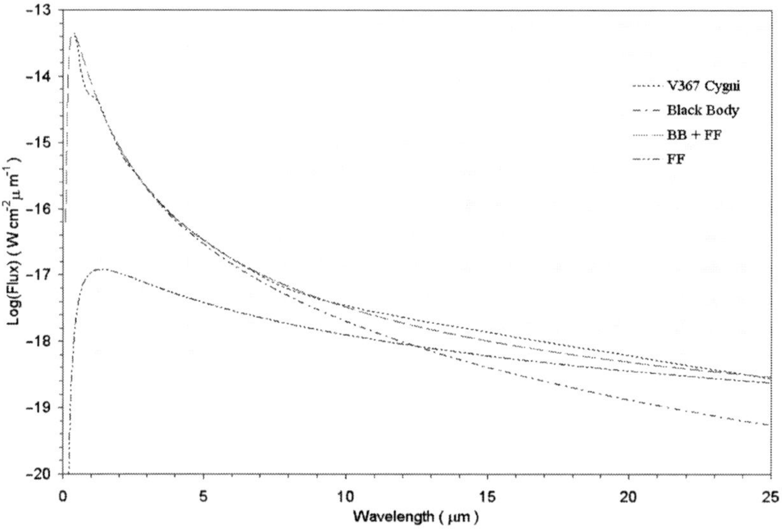

Fig. 1 Spectral energy distribution of V367 Cygni

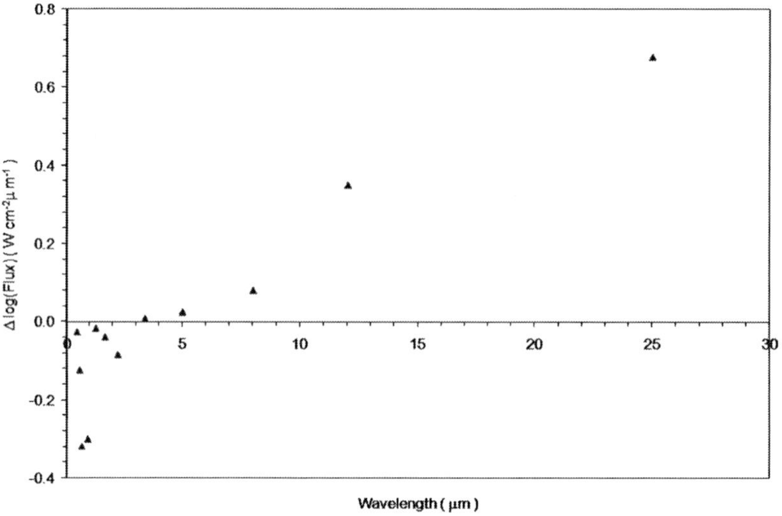

Fig. 2 Near infrared excess of V367 Cygni

was fitted with the SED of V367 Cygni by normalizing the two curves at λ 4,400. We found that the two curves fit fairly well up to 3 μm. An increasing near infrared excess flux is observed from the binary system beyond 3 μm. (Fig. 2). Free–free emission from the ionized accretion disk is attributed to this infrared excess flux as described in the following.

The free–free volume emission coefficient is given by (Banerjee et al. 2001)

$$j_{\lambda_{ff}} = 2.05 \times 10^{-30} \lambda^{-2} Z^2 g T^{-0.5} n_e n_i e^{(-c_2/\lambda T)} \text{ Wcm}^{-3}\, \mu\text{m}^{-1} \quad \rightarrow \quad (2)$$

Where λ is the wavelength of emission in μm, z is the charge, g is the Gaunt factor, T is the temperature, n_e and n_i are the electron and ion densities respectively and $c_2 = 1.438\,\mu$m. The free–free contribution from the circumstellar disk can be obtained by multiplying the volume emission coefficient by the volume of the disk. In this study we consider a thick flat disk whose volume V_d is given by

$$v_d = \pi \left(r_s^2 - r_d^2\right) z_d$$

where r_d is the radius of the disk, z_d is the thickness of the disk and r_s is the radius of the secondary component.

The observed flux is given by $F_{\lambda_{ff}} = \frac{j_{\lambda_{ff}} v_d}{4\pi D^z}$

$$F_{\lambda_{ff}} = \frac{j_{\lambda_{ff}} \left(r_s^2 - r_d^2\right) z_d}{4D^2} \quad \rightarrow \quad (3)$$

Where D is the distance to the binary system.

In the above equation for $j_{\lambda ff}$ we use g = 1 and assume that the disk is pure hydrogen. The charge z is taken as 1 and $n_i = n_e$.

We consider the parameters of the binary system quoted by K Pavlovski and S Zola in their light curve modeling separately in our calculations. The stellar parameters of the binary system claimed by these two authors are given in the Table 3.

We developed a computer code to generate the free–free emission observed flux using (2) and (3). This observed free–free flux is added to the black body flux and tried to find a good fit for the observed excess flux, leaving n_e as a free parameter. See the Fig. 2.

In our observed free–free emission flux calculations, we used the parallax of V367 Cygni, 0.88 mas found by the Hipparcos to calculate the distance to the system. As the uncertainty of the measurement is ±0.63 mas we looked for possible values of n_e at both extreme limits of the uncertainty. The calculated values of electron density(n_e) which are found to give a good fit and the volume emission measure(n_e^2v) are given in the Table 4.

Taranova has found the volume emission measure of the excess emission is $\sim 8 \times 10^{59}$ cm^{-3}. It can be seen from the Table 4 that the values we obtained using

Table 3 Absolute parameters of the V367 Cygni

	K Pavlovski	S Zola
T_d	6,575° K	5,300° K
r_s	4.98 R_O	2.3 R_O
r_d	23.78 R_O	17.9 R_O
z_d	2.27 R_O	5.1 R_O
r_d/z_d	10.48	3.5

Table 4 Electron density and volume emission of the ionized disk

	K Pavlovski		S Zola	
Parallax mas	$n_e cm^{-3}$	$n_e^2 v\ cm^{-3}$	$n_e cm^{-3}$	$n_e^2 v\ cm^{-3}$
0.25	3.0×10^{12}	1.17×10^{61}	1.8×10^{12}	1.2×10^{61}
0.88	9.0×10^{11}	1.05×10^{60}	5.0×10^{11}	9.1×10^{59}
1.51	5.0×10^{11}	3.25×10^{59}	3.0×10^{11}	3.3×10^{59}

the stellar parameters derived by S Zola et al. are in good agreement within the uncertainties of the measurement of the parallax of the system.

Plavec (1980) has found in his study on W Serpentis stars that these stars have surprisingly high electron densities (β Lyrae $\sim 3 \times 10^{12}$ cm^{-3}.).

3 Discussion

We obtained spectral energy distribution of V367 Cygni using previous photometric measurements done on this system in different spectral bands. We fitted a black body curve of 8,000 K to this system and found that there is an infrared excess flux starting from 3 μm. Free–free emission is attributed to this excess energy and the electron density of the mass accretion disk of the system is derived by fitting the free–free emission to the observed excess flux.

Volume emission measure of the circumstellar disk is found to be $\sim 9 \times 10^{59}$ cm^{-3} and is in good agreement with the values found by Taranova and Plavec (1980) within the uncertainties in the measurements of the parallax of the system.

References

S. Zola and W. Ogloza, A&A 368, 932–938 (2001)
D. P. K. Banerjee, P. Janardhan, and N. M. Ashok, A&A 380, L13–L16 (2001)
K. Pavlovski, H. Schneider, and M. C. Akan, A&A 258, 329–334 (1992)
M. J. Plavec, IAU (1980)

Period Study and Secondary Maximum of KZ Hya

Fredy Doncel and Takashi Momiyama

Abstract We have carried out CCD photometry of KZ Hya from June 2004 to April 2007. With the 45 cm reflecting telescope donated by Japanese Official Development Assistance, we obtained 20 light maxima at the Astronomical Observatory of the National University of Asuncion in Paraguay. In addition, we have partly obtained three light maxima at the Gunma Astronomical Observatory in Japan. We have collected additional maximum data from literatures and data bases. The $O - C$ values of KZ Hya varies sinusoidally with a period of 7.96 years, which supports the binary nature of KZ Hya suggested by some other workers. We have also clearly detected the secondary maximum of 0.03 mag at the phase of 0.66 on the light curve. The secondary maximum is found to have been present at the same phase since 1975 when KZ Hya was discovered as a pulsating variable star.

Keywords Stars · Oscillations · Binaries · Ephemerides · Individual (KZ Hya)

1 Introduction

KZ Hya ($V = 9.46 - 10.26$, $P = 0.0595104212$ days according to GCVS2000) is a SX Phe type pulsating star, whose $O - C$ values exhibit a long period regular variation interpreted as the light time effect of the binary system described by several authors. Liu et al. (1991) and Bonnardeau (2007) have derived the orbital period as $P = 33829.5$ days (9.262 years) and 7.934 years, respectively.

Several authors have mentioned the bump or secondary maximum on the light curve of KZ Hya. Przybylski and Bessell (1979) mentioned the secondary maximum of about 0.03 mag at the phase 0.76 on the light curve. Doncel et al. (2004) found the marginal secondary maximum around the phase 0.7. Bonnardeau (2007) has mentioned that the bump appears around the phase 0.7. Napoleao (2003) also

F. Doncel (✉) and T. Momiyama
Astronomical Observatory FP-UNA, Campus Universitario de San Lorenzo, Paraguay
e-mail: fdoncel@pol.una.py

H.J. Haubold and A.M. Mathai (eds.), *Proceedings of the Third UN/ESA/NASA Workshop on the International Heliophysical Year 2007 and Basic Space Science*, Astrophysics and Space Science Proceedings, DOI 10.1007/978-3-642-03325-4_15,

detected the secondary max of 0.03 mag. However, none of these authors have made any discussion on the secondary maximum on the basis of all the available light curves since the discovery.

In order to study the binary nature and the secondary maximum in more detail, we have made photometric monitoring of KZ Hya.

2 Observations and Data Collection

We obtained 20 new light maxima for KZ Hya from June 2004 to July 2006 using the 45 cm Cassegrain telescope of f/12 at the Astronomical Observatory of the National University of Asuncion in Paraguay which was donated by Japanese Official Development (ODA). Observations were made in the V, R, and I bands using an ST-8 (SBIG) CCD camera providing an 8'.7 × 5'.8 field of view. The integration time ranged between 10 and 40 s according to the quality of night. In addition, we have also obtained three new times of maximum light in April 2007 using the 65 cm Cassegrain telescope of f/12 at the Gunma Astronomical Observatory in Japan. The observations have been made in the R band using an AP8 (Apogee) CCD camera providing a 10'.6 × 10'.6 field of view. The integration time was 15 s. For each observation, HD 93998 (= TYC 6638-764-1) was chosen as the comparison star. This is the same star used by Doncel et al. (2004). The MaxIm DL (Diffraction Limited) has been used for the acquisition of data for each observatory and all photometric reductions have been performed using the AIP4Win (Willmann-Bell, Inc.) and some Excel macros developed by the authors. In total, we have obtained 23 light maxima in our observations.

3 Analysis for the Period Change

To find the new ephemeris, we have collected light maximum data which have been reported so far. For example, we collected 25 light maxima from Przybylski and Bessell (1979), and 38 maxima from Hobart et al. (1985). Four maxima from McNamara and Budge (1985) were examined and one data (HJD2443604.4488) were revised to be HJD2443604.6688 by Doncel et al. (2004) using their photometry data (Table 1 of the paper). For using the data from Liu et al. (1991), we did not include three data (HJD2446153.0652, 2446153.1298, and 2448398.1510) listed in the table of their paper, which are discrepant between $O - C$ values in the Table 1 and their plotted positions in Fig. 1 in their paper. Then we adopted 21 times of light maximum from Liu et al. (1991). From ASAS (2005), three maxima were obtained by folding the data with the period 0.05911157 (given by Doncel et al. (2004)) over each season. From the R band data of AAVSO (2004), we obtained one maximum. Two maxima were obtained form VSOLJ (1998). 14 maxima from Bonnardeau (2007) have been collected. Napoleao (2003) observed V and R band simultaneously using two telescopes and obtained six maxima for each band. We used the V

Table 1 Times of maximum light and $O-C$ residuals of KZ Hya

No.	Time of Maximum (HJD)	Epoch	$O-C$ (d)	Residual (d)	Ref.
1	2442516.1585	0	0.0001	0.0005	1
2	2442517.9440	30	0.0003	0.0006	1
3	2442518.0034	31	0.0002	0.0006	1
4	2442518.1223	33	0.0001	0.0005	1
5	2442541.9266	433	−0.0002	0.0002	1
6	2442541.9859	434	−0.0003	0.0000	1
7	2442542.0448	435	−0.0010	−0.0007	1
8	2442542.1045	436	−0.0008	−0.0005	1
9	2442542.8787	449	−0.0002	0.0002	1
10	2442542.9382	450	−0.0002	0.0001	1
11	2442542.9973	451	−0.0006	−0.0003	1
12	2442543.8903	466	−0.0003	0.0001	1
13	2442544.0094	468	−0.0002	0.0001	1
14	2442545.9138	500	−0.0002	0.0001	1
15	2442545.9729	501	−0.0006	−0.0003	1
16	2442562.9930	787	−0.0007	−0.0003	1
17	2442755.2111	4017	−0.0040	−0.0029	1
18	2442846.9767	5559	−0.0047	−0.0028	1
19	2442847.0959	5561	−0.0046	−0.0027	1
20	2442847.1550	5562	−0.0049	−0.0031	1
21	2442890.0628	6283	−0.0047	−0.0025	1
22	2442890.9553	6298	−0.0050	−0.0027	1
23	2442891.9664	6315	−0.0055	−0.0032	1
24	2442929.9939	6954	−0.0057	−0.0030	1
25	2442958.9161	7440	−0.0060	−0.0030	1
26	2443601.6931	18241	−0.0100	−0.0008	2
27	2443604.6688	18291	−0.0099	−0.0007	2
28	2444664.2720	36096	−0.0046	0.0010	3
29	2444690.0997	36530	−0.0048	0.0005	3
30	2444690.1593	36531	−0.0047	0.0006	3
31	2444691.1109	36547	−0.0053	0.0000	3
32	2444691.1705	36548	−0.0052	0.0001	3
33	2445383.2904	48178	−0.0012	−0.0006	3
34	2445384.1834	48193	−0.0008	−0.0002	3
35	2445384.2430	48194	−0.0007	−0.0001	3
36	2445748.7479	54319	−0.0023	−0.0001	4
37	2445748.8081	54320	−0.0016	0.0005	4
38	2445748.8671	54321	−0.0021	0.0000	4
39	2445748.9278	54322	−0.0009	0.0012	4
40	2445748.9861	54323	−0.0021	0.0000	4
41	2445749.7608	54336	−0.0011	0.0011	4
42	2445749.8199	54337	−0.0015	0.0007	4
43	2445749.8795	54338	−0.0014	0.0007	4
44	2445749.9387	54339	−0.0017	0.0004	4
45	2445750.9503	54356	−0.0018	0.0004	4
46	2445751.7243	54369	−0.0014	0.0007	4

(continued)

Table 1 (continued)

No.	Time of Maximum (HJD)	Epoch	$O-C$ (d)	Residual (d)	Ref.
47	2445769.7564	54672	−0.0012	0.0011	4
48	2445769.8158	54673	−0.0013	0.0010	4
49	2445769.8750	54674	−0.0017	0.0007	4
50	2445770.6479	54687	−0.0024	−0.0001	4
51	2445770.7077	54688	−0.0021	0.0002	4
52	2445770.7674	54689	−0.0019	0.0004	4
53	2445776.7185	54789	−0.0020	0.0004	4
54	2445777.6699	54805	−0.0027	−0.0003	4
55	2445777.7303	54806	−0.0018	0.0005	4
56	2445782.6699	54889	−0.0017	0.0008	4
57	2445782.7294	54890	−0.0017	0.0007	4
58	2445782.7890	54891	−0.0016	0.0008	4
59	2445782.8485	54892	−0.0016	0.0008	4
60	2445783.6812	54906	−0.0021	0.0004	4
61	2445783.7410	54907	−0.0018	0.0007	4
62	2445783.8000	54908	−0.0023	0.0002	4
63	2445784.6935	54923	−0.0015	0.0010	4
64	2445784.7527	54924	−0.0018	0.0007	4
65	2445784.8125	54925	−0.0015	0.0010	4
66	2445792.6660	55057	−0.0035	−0.0009	4
67	2445793.2037	55066	−0.0014	0.0012	3
68	2445793.2629	55067	−0.0017	0.0009	3
69	2445793.6792	55074	−0.0020	0.0006	4
70	2445795.0488	55097	−0.0011	0.0014	3
71	2445795.1073	55098	−0.0021	0.0004	3
72	2445795.1670	55099	−0.0019	0.0006	3
73	2445795.6431	55107	−0.0019	0.0006	4
74	2445796.0606	55114	−0.0010	0.0016	3
75	2445796.1194	55115	−0.0017	0.0008	3
76	2445796.1793	55116	−0.0013	0.0012	3
77	2445798.6785	55158	−0.0016	0.0010	4
78	2445800.6422	55191	−0.0018	0.0008	4
79	2445808.6168	55325	−0.0017	0.0010	4
80	2445808.6761	55326	−0.0019	0.0008	4
81	2445808.7358	55327	−0.0017	0.0010	4
82	2445854.4989	56096	−0.0028	0.0004	2
83	2445856.4612	56129	−0.0043	−0.0011	2
84	2446501.2014	66963	−0.0091	0.0005	3
85	2446503.1650	66996	−0.0093	0.0003	3
86	2446503.2248	66997	−0.0090	0.0006	3
87	2446889.1548	73482	−0.0095	0.0007	3
88	2451175.2830	145504	−0.0009	0.0016	5
89	2451175.3430	145505	−0.0004	0.0021	5
90	2452383.5289	165807	−0.0119	0.0002	6
91	2452387.5160	165874	−0.0121	0.0000	6
92	2452387.5749	165875	−0.0127	−0.0006	6

(continued)

Table 1 (continued)

No.	Time of Maximum (HJD)	Epoch	$O-C$ (d)	Residual (d)	Ref.
93	2452403.5242	166143	−0.0124	−0.0002	6
94	2452403.5835	166144	−0.0126	−0.0004	6
95	2452404.5359	166160	−0.0124	−0.0002	6
96	2452445.4795	166848	−0.0126	−0.0002	6
97	2452464.4630	167167	−0.0131	−0.0007	6
98	2452473.4498	167318	−0.0125	−0.0001	6
99	2452723.4568	171519	−0.0124	0.0001	7
100	2452732.5624	171672	−0.0119	0.0006	8
101	2452732.6222	171673	−0.0116	0.0008	8
102	2452732.6813	171674	−0.0120	0.0004	8
103	2452732.7412	171675	−0.0117	0.0008	8
104	2452732.8006	171676	−0.0118	0.0007	8
105	2452732.8599	171677	−0.0120	0.0004	8
106	2453090.8799	177693	−0.0117	−0.0008	7
107	2453144.3804	178592	−0.0118	−0.0013	9
108	2453168.485	178997	−0.0097	0.0007	11
109	2453462.6504	183940	−0.0080	0.0000	7
110	2453516.509	184845	−0.0070	0.0006	11
111	2453516.5681	184846	−0.0075	0.0000	11
112	2453516.6276	184847	−0.0075	0.0000	11
113	2453523.4717	184962	−0.0072	0.0003	11
114	2453523.5908	184964	−0.0071	0.0004	11
115	2453524.4238	184978	−0.0073	0.0002	11
116	2453524.4826	184979	−0.0080	−0.0005	11
117	2453525.4355	184995	−0.0073	0.0002	11
118	2453525.4948	184996	−0.0075	0.0000	11
119	2453525.5544	184997	−0.0074	0.0001	11
120	2453542.4554	185281	−0.0076	−0.0003	11
121	2453542.5155	185282	−0.0070	0.0003	11
122	2453542.5753	185283	−0.0067	0.0006	11
123	2453557.5116	185534	−0.0077	−0.0006	11
124	2453558.4644	185550	−0.0071	0.0001	11
125	2453570.4857	185752	−0.0071	0.0000	11
126	2453728.7275	188411	−0.0057	0.0001	10
127	2453729.6795	188427	−0.0059	−0.0001	10
128	2453761.5775	188963	−0.0059	−0.0004	10
129	2453761.6375	188964	−0.0054	0.0001	10
130	2453808.4135	189750	−0.0053	−0.0001	10
131	2453808.4730	189751	−0.0053	−0.0001	10
132	2453808.5320	189752	−0.0058	−0.0006	10
133	2453833.4080	190170	−0.0055	−0.0005	10
134	2453906.4883	191398	−0.0050	−0.0004	11
135	2453916.4864	191566	−0.0048	−0.0002	11
136	2453920.5331	191634	−0.0049	−0.0003	11
137	2454141.5583	195348	−0.0045	−0.0005	10
138	2454171.4927	195851	−0.0043	−0.0002	10

(continued)

Table 1 (continued)

No.	Time of Maximum (HJD)	Epoch	$O-C$ (d)	Residual (d)	Ref.
139	2454174.4090	195900	−0.0040	0.0001	10
140	2454174.4680	195901	−0.0045	−0.0005	10
141	2454174.5275	195902	−0.0045	−0.0005	10
142	2454210.4127	196505	−0.0046	−0.0005	10
143	2454216.0066	196599	−0.0048	−0.0006	11
144	2454216.0659	196600	−0.0050	−0.0008	11
145	2454217.0189	196616	−0.0042	0.0000	11

References. — (1) Przybylski and Bessell (1979); (2) McNamara and Budge (1985); (3) Liu et al. (1991); (4) Hobart et al. (1985); (5) VSOLJ (1998); (6) Doncel et al. (2004). Times of maximum were re-examined by the authors. See text. (7) ASAS (2005) (8) Napoleao (2003); (9) AAVSO (2004); (10) Bonnardeau (2007); (11) Present study.

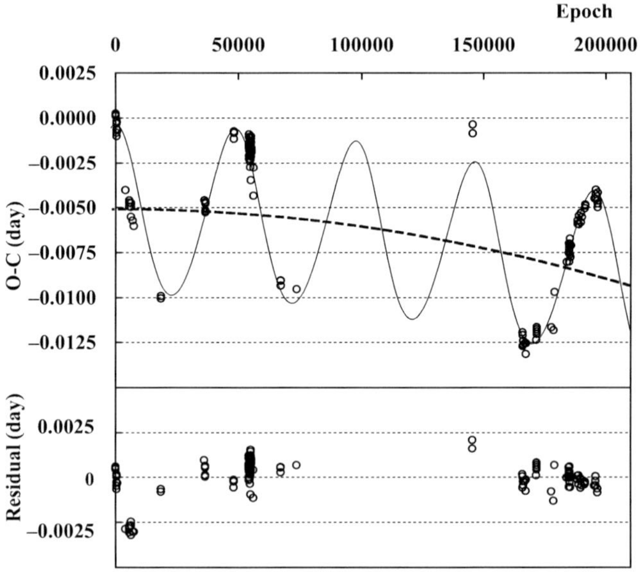

Fig. 1 The top panel shows $O-C$ diagram and bottom panel shows the residual of KZ Hya. In the top panel, the open circles are $O-C$ value computed by our linear ephemeris (1). The broken line shows the quadratic ephemeris and the solid line represents the best fit ephemeris with the quadratic and sinusoidal term (2). Their pulsation and orbital elements are shown in Table 2. In the bottom panel, the open circles show the residuals of $O-C$ value from the quadratic and sinusoidal curve

band data in the present study. Doncel et al. (2004) obtained 12 maxima with the same instruments as ours in Paraguay, which have turned out to have some problems in the time calibration. We have re-examined them and have determined 9 maxima.

Finally, 145 times of light maxima of KZ Hya are now available to us.

Table 1 gives all the times of maximum light used to study $O-C$ residuals of KZ Hya. Column 1 gives the data number, column 2 gives the time of maximum light observed, column 3 gives the epoch (cycle number) of each maximum, column 4 gives $O - C$ value computed with our linear ephemeris given by the following equation;

$$\begin{aligned} T_{max} &= T_0 + P_0 \cdot E \\ &= \text{HJD}2442516.15836 + 0.059511257 \cdot E, \end{aligned} \tag{1}$$

where T_{max} is the computed time of maximum light, T_0 is the initial epoch, P_0 is the period and E is the epoch (cycle number).

Figure 1 (top panel) shows the $O - C$ diagram, where the data in column 4 of Table 1 are used. In the $O - C$ diagram, the data are fitted to a quadratic and sinusoidal curve, which corresponds to the following equation;

$$T_{max} = T_0 + P_0 \cdot E + 0.5 \cdot \beta \cdot E^2 + \Delta T, \tag{2}$$

where the term $0.5 \cdot \beta \cdot E^2$ represents the secular variation and ΔT is the light time effect of the binary system, which is determined by Irwin (1952);

$$\Delta T = \frac{a \cdot \sin(i)}{c} \cdot \left\{ \frac{1 - e^2}{1 + e \cdot \cos(\nu)} \cdot \sin(\nu + \omega) + e \cdot \sin(\omega) \right\} \tag{3}$$

where

$$\nu = 2 \cdot \tan^{-1} \left\{ \sqrt{\frac{1 + e}{1 - e}} \cdot \tan \frac{\phi}{2} \right\} . \tag{4}$$

In (3) and (4), $a \cdot \sin(i)$ is a projected semi major axis of the orbit, c is the velocity of light, e is the eccentricity of the orbit, ν is the true anomaly, ω is the longitude of the periastron passage and ϕ is the eccentric anomaly.

The best fit curve, deduced by the least square method, is also drawn with the solid line in Fig. 1 (top panel), where the curve corresponding to the quadratic term is also shown with the broken line. The pulsation and orbital elements, which gives the best fit curve, are shown in Table 2 together with Liu et al. (1991) and Bonnardeau (2007) for comparison. Column 5 of Table 1 gives residuals from the best fit curve, which is also plotted in Fig. 1 (bottom panel). The standard deviation of the residual is 0.0009 days. The smallness of residuals supports the binary nature of KZ Hya as previously suggested by Liu et al. (1991) and Bonnardeau (2007). We found the secular period change, the binary period and the eccentricity of the orbit of KZ Hya to be -5.38×10^{-14} days·cycle^{-1}, 7.96 years and 0.06, respectively. The eccentricities reported in the previous studies are larger than that obtained in the present study. Newly collected data are densely distributed around the recent ascending phase of the sinusoidal curve, and $O - C$ value is turning into descending phase around 2007, so continuous observation has to be needed to confirm the ephemeris.

Table 2 Pulsation and orbital elements

Element		Liu et al. (1991)	Bonnardeau (2007)	Present study
Starting epoch (HJD)	T_0	2442516.15576	2442516.15584	2442516.15836
Period (day)	P_0	0.059511036	0.059511264	0.059511257
Period change (days · cycle^{-1})	β	2.92×10^{-12}	- - -	-5.38×10^{-14}
Binary period (years)	P_b	33829.5 days (= 9.262 years)	7.934	7.96
Projected semi major axis (au)	$a \cdot \sin(i)$	0.95	0.889	0.82
Eccentricity	e	0.20	0.445	0.06
Time of periastron passage (HJD)	τ	2442708.4	2442895.9	2437032.0
Longitude of the periastron passage (deg)	ω	175	153*	142

*In his study, Bonnardeau (2007) gives 333 deg, which corresponds to 153 deg in our definition.

4 Secondary maximum

We have clearly obtained the secondary maximum on the light curve of KZ Hya. Figure 2 shows the secondary maximum observed on 3 June 2005 compared with previous observation by Doncel et al. (2004). In our observation, two shallow minima are seen on both side of the secondary maximum. We call the foregoing minimum as min1 and the following minimum as min2, respectively, in the present study. We have obtained 15 min1s, 13 secondary maxima and 14 min2s from our observations and have found that their phases are 0.55 ± 0.02, 0.66 ± 0.02, and 0.74 ± 0.02, respectively. We also have found that secondary max is brighter than min1 by 0.03 ± 0.01 mag.

Figure 3 shows five light curves which include two of our observations ((4), (5)) and three from other sources ((1), (2), (3)), which covers the period from 1978 to 2006. In Fig. 3, light curves of McNamara and Budge (1985), VSOLJ (1998), and ASAS (2005) are not so clear to determine the phases concerned with the secondary maximum, but we tried to estimate the phase of them from the smoothed mean light curve. We have found that min1s are around the phase 0.52, 0.57, and 0.55, respectively, and the secondary maxima are around 0.62, 0.67, and 0.66, respectively. Min2 has been estimated only for ASAS (2005) around 0.73. The magnitude differences between the secondary maximum and min1 have been estimated as 0.03 mag, 0.04 mag, and 0.03 mag, respectively. Przybylski and Bessell (1979) mentioned from their observation during 1975–1976 that the star reaches minimum brightness at the phase 0.58 and the slow rise after this minimum was followed by a secondary drop in brightness by about 0.03 mag at the phase 0.76.

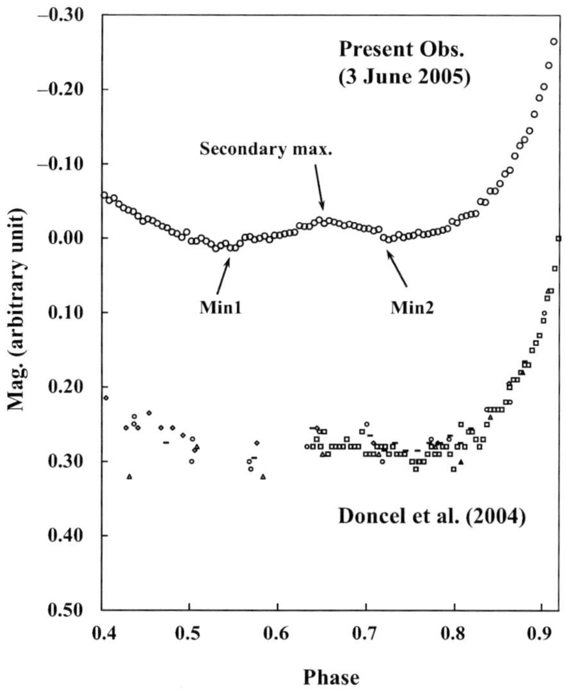

Fig. 2 Secondary maximum was observed clearly in this study. The upper light curve represents the *I* band observation on 3 June 2005 and lower one represents Doncel et al. (2004)

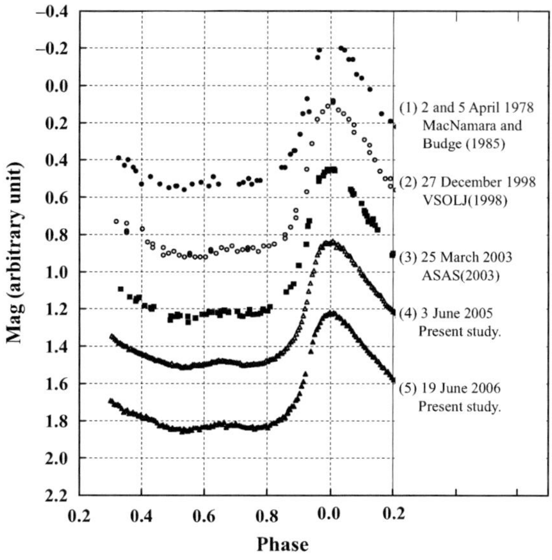

Fig. 3 Five light curves observed in different years are shown with date (UT) and their sources. From the top, (1) the data of two days from McNamara and Budge (1985) were merged. (2) HJD at phase 0.0 is 2451175.283. (3) The data of ASAS from HJD2452622 to HJD2452845 were folded by the period 0.05911157. Phase 0.0 corresponds to HJD2452723.4568. (4) Phase 0.0 corresponds to HJD2453525.4948. (5) Phase 0.0 corresponds to HJD2453906.4883

All of the above data concerned with the secondary maximum suggest that it has been present at the same phase substantially since 1975 when KZ Hya was first discovered as a pulsating variable star. As far as we know, KZ Hya is the only SX Phe type variable which has always shown stable secondary maximum, though Blake et al. (2003) marginally detected a possible bump on the light curve of XX Cyg of SX Phe type star to be a transient feature.

5 Conclusion

We have obtained 23 light maxima in our CCD photometric observations of KZ Hya. The theoretical curve of the light time effect of the binary system has well fitted the $O-C$ values, which has the binary period of 7.96 years. It supports the binary nature of KZ Hya as previously suggested by Liu et al. (1991) and Bonnardeau (2007). We have clearly detected the existence of the secondary maximum of 0.03 mag at the phase of 0.66 on the light curve. The data concerned with the secondary maximum suggest that it has been present at the same phase substantially since 1975 when KZ Hya was discovered as a pulsating variable star.

Acknowledgements We should like to express our gratitude to Prof. M. Kitamura of the National Astronomical Observatory of Japan, Prof. A. Okazaki of Gunma University in Japan and Associate Prof. K. Yoshioka of the University of the Air in Japan for their helpful discussions. We acknowledge with thanks the variable star observations from the AAVSO International Database contributed by observers worldwide and used in this research. The variable star observations from the VSOLJ Japanese Database are acknowledged.

References

American Association of Variable Star Observers: International Database. http://www.aavso.org/. Cited 27 Feb 2007(2004)

The All Sky Automated Survey Data. The ASAS-3 Catalog of Variable Stars. http://archive.princeton.edu/asas/. Cited 16 Mar 2007(2005)

Blake, R. M., Delaney, P., Khosravani, H., Tome, J. and Lightman, M.: PASP **115**, 212(2003)

Bonnardeau, M.: http://mbond.free.fr/. Cited 14 May 2007(2007)

Doncel, F., Troche, A. and Noguchi, T.: UNSPA **15**, 29(2004)

Hobart, M. A., Peniche, R. And Pena, J. H.: Rev. Mex. Astron. Astrof. **11**, 19(1985)

Irwin, J. B.: ApJ. **116**, 211(1952)

Liu Yanying, Jiang Shiyang and Cao Ming: IBVS No.3606(1991)

McNamara, D. H. and Budge, K. G.: PASP **97**, 322(1985)

Napoleao, T. A.: REA/Brasil REPORTE No.11(2003)

Pojmanski, G.: Acta Astronomica **52**, 397(2002)

Przybylski, A. and Bessell, M.S.: Mon. Not. R. astr. Soc. **189**, 377(1979)

Variable Star Observers League in Japan: Database. ftp://ftp.kusastro.kyoto-u.ac.jp/pub/vsnet/VSOLJ/database/stars/HYAKZ.ut. Cited 11 Feb 2007(1998)

Current and Future Capabilities of the 74-inch Telescope of Kottamia Astronomical Observatory in Egypt

Y.A. Azzam, G.B. Ali, F. Elnagahy, H.A. Ismail, A. Haroon, and I. Selim

Abstract In this paper, we are going to introduce the Kottamia Astronomical Observatory, KAO, to the astronomical community. The current status of the telescope together with the available instrumentations is described. An upgrade stage including a new optical system and a computer controlling of both the telescope and dome are achieved. The specifications of a set of CCD cameras for direct imaging and spectroscopy are given. A grating spectrograph is recently gifted to KAO from Okayama Astrophysical Observatory, OAO, of the National Astronomical Observatories in Japan. This spectrograph is successfully tested and installed at the F/18 Cassegrain focus of the KAO 74″ telescope.

1 Introduction

In 1905, Mr. Reynolds, an amateur astronomer at that time and later treasurer of the Royal Astronomical Society in London presented Helwan Observatory with a 30″ reflecting telescope. Due to the clear sky of Helwan and the ability of the astronomers in charge of this telescope, a great deal of valuable observations was collected. As of the curiosity of astronomers to achieve more, a large telescope was necessary and a 74″ telescope equipped with both Cassegrain and Coude spectrographs had been recommended to the Egyptian authorities. The Egyptian government signed a contract for this purpose with Messers. Grubb Parsons of Newcastle. U.K. in the same year (1948) that the giant 200″ telescope of Mount Palomar California was erected. The delivery of the telescope and spectrographs was expected in 1955, but various difficulties arose which resulted in a considerable delay of delivery until 1963 (Samaha 1964).

Kottamia mountain (about 80 Kms. NE of Helwan, 12 km. north of mid-point of Cairo-Ein Sukhna road, 476 m above see-level) had been chosen for the location

Y.A. Azzam (✉), G.B. Ali, F. Elnagahy, H.A. Ismail, A. Haroon, and I. Selim
National Research Institute of Astronomy and Geophysics, Helwan, Cairo, Egypt
e-mail: azzam@nriag.sci.eg

H.J. Haubold and A.M. Mathai (eds.), *Proceedings of the Third UN/ESA/NASA Workshop on the International Heliophysical Year 2007 and Basic Space Science*, Astrophysics and Space Science Proceedings, DOI 10.1007/978-3-642-03325-4_16,

of Kottamia Astronomical observatory (KAO). The latitude of the observatory is 29°55′ 35.24″ N; the longitude is 31°49′ 45.85″ E. The seeing conditions prevailing at the site is around 2 arcsec. in average (Hassan 1998). KAO belongs to the National Research Institute of Astronomy and Geophysics; NRIAG (previously, Helwan Observatory).

2 General Overview of the Kottamia 74″ Telescope

The reflector is of conventional design, having a parabolic mirror with secondaries to give Newtonian, Cassegrain and Coudé foci. The mounting is of English type with north and south piers that support a polar axis with the tube offset on a short declination axis (Issa 1986). A photo in Fig. 1 shows the Kottamia telescope with its Cassegrain and Newtonian foci.

The main mirror has an aperture of 74 in. and a paraboloidal surface of 360″ focal length (F/4.9). The Cassegrain secondary has an aperture of 19″ and a hyperbolic section giving a focal ratio of F/18, with plate scale of 6.1 arcsec mm^{-1}.

A second hyperbolic mirror similar to the Cassegrain has a focal ratio of F/28.9 and in combination with two flats of 13.5 and 9.5 in. apertures give the Coude focus plate scale of 3.8 arcsec mm^{-1}.

Fig. 1 Kottamia 74" Telescope

The Newtonian focus has a focal ratio of F/4.9, thus giving a scale of 22.53 arcsec mm^{-1}. A review of the values of focal lengths, F/ratios and plate scales at various foci of the telescope is given in Table 1.

3 Telescope Dome

The dome was constructed by the united Austrian Iron and Steel Works (VOEST) of Linz/Austria. It is 18 m in diameter with an outer shell of galvanized steel plate painted externally with aluminum. The dome is very good insulated from heat and dust which lead to diurnal range of temperature variation to be as minimum as 3°C. A photo showing the dome is given in Fig. 2.

The dome can be rotated at a speed of one revolution in 6 min by means of six three phase motors driving pinions into a circular rack. The dome runs on 26 traveling wheels and is kept central by means of 15 horizontal guide rollers placed around the circumference to allow for thermal expansion and distributions of wind loading. The dome has two shutters, an upper and a lower capable of moving at 1.5 m min^{-1}. Hand operation is possible in case of emergency. The upper shutter is

Table 1 The Parameters of the KAO's Foci

Focus	Focal length (m)	F-number	Plate scale (arcsec mm^{-1})
Newtonian	9.15	4.9	22.53
Cassegrain	33.98	18	6.09
Coude	54.29	28.9	3.8

Fig. 2 The KAO Dome

of the "up and over" type whereas the lower shutter is in two parts each opening to the side. A canvas wind screen provide the protective action of the lower shutter. The opening of the dome is 5 m wide and extends 2.5 m beyond the Zenith. An observing carriage traveling at 4 m min^{-1} is fitted with control buttons for dome rotation, shutter operations and its own motion. It runs on rails attached to the main arches of the dome. Similar control pushbuttons are also provided at various positions around the periphery of the dome. At present there are no electrical control connections between the telescope and the dome.

Some few years ago, the dome control has been upgraded by one of us (F. Elnagahy) to an RF handset control. As the telescope tracks the object across the sky, it is necessary to rotate the dome from time to time to allow the light of the object to enter to the telescope. This is being done recently by the RF handset control instead of pushbuttons around the periphery of the dome, which previously necessitated a person to go up the stairs to do such movement action.

4 Kottamia Telescope Upgrading

In order to share astronomical communities their enormous contributions in astronomical research, it was necessary to modernize an approximately 32-years old 74″ telescope. The matter was raised to governmental authorities and after some long subjective discussions, it was approved to support Kottamia telescope with the necessary money for upgrading.

4.1 Optical System Upgrade

In 1995, an appropriate agreement was signed with Zeiss company (FRG) to design a new optical system for the telescope. This involved a new primary mirror, secondary Cassegrain mirror both are made of "Zerodur" glass ceramics, and a new supporting system for the primary mirror (Hassan 2004). Due to some complications and problems in the design of the supporting system, the system was not approved until 2003 when it was being in effect and working properly.

The technical parameters of such optical system is as follows:

Primary mirror

Diameter	1,930 mm
Diameter of center hole	188 mm
Thickness on the edge	230 mm
Optical aperture	1,880 mm
Radius of curvature	$-18,277$ mm
Material	ZERODUR
Thermal expansion coefficient	$0 \pm 0.1 \times 10^{-6}/\text{K}$

Secondary mirror

Diameter	496.8 mm
Thickness in center	80.1 mm
Optical aperture	483 mm
Radius of curvature	5877.62 mm
Conical constant	$k = -3.012377$
Material	ZERODUR
Thermal expansion coefficient	$0 \pm 0.1 \times 10^{-6}/K$

Optical system

Focal ratio – Focal length	F/18–33,984 mm
Distance M1 M2 vertex distance	6989.9 mm
M1 vertex – focal plane	1,000 mm

Primary mirror support

Number of axial supports	18
Number of radial supports	16
Distance focal plane-instrument rotator flange	417.5 mm
Maximum load onto instrument rotator flange	400 kg with 1,000 mm distance to center of gravity

4.2 CCD Camera System

Before the time of the acceptance of the optical system and in 1997, the telescope has been equipped with CCD imaging system for direct imaging. The system was provided by AstroCam (later PixCellent) in U.K. This CCD system consists of two cooled CCD systems built onto an assembly that allows mounting at either the Newtonian or the Cassegrain foci of the telescope. The system has a primary CCD system consisting of a SITe (formerly Tektronics) 1,024 × 1,024 pixel (24 μm per pixel pitch) back illuminated CCD mounted in a liquid nitrogen vacuum dewar and controlled by an AstroCam 4202 controller with the Imager 2 software package. In addition, there is a second cooled CCD system that is thermoelectrically cooled. It uses a Kodak KAF-1300 CCD of 1,280 × 1,024 pixels each of 16 μm. This system is also controlled by an AstroCam 4202 CCD controller and Imager 2 software. Its purpose is to act as an offset acquisition and guiding system to allow manual operation of the telescope to set it accurately to follow the science field of view without any image trailing (Osman 2000). A photo showing the camera system is shown in Fig. 3, where it is attached to the Cassegrain focus of the telescope, and detailed specifications of both cameras are given in Tables 2 and 3.

Fig. 3 Kottamia CCD camera system attached to the Casegrain focus

Table 2 Primary CCD camera system specifications

Type	SITe (formerly Tektronics)
Format	10, 241 pixels × 1, 024 pixels full frame
Pixel size	24 μm × 24 μm
Imaging area	24.6 mm × 24.6 mm
Dark current @20°C	0.1 nA cm^{-2}
Readout noise	7 electrons
Full well signal	150.000 electrons
Dynamic range	20.000–30.000:1
Quantum efficiency at room temperature	
λ = 400, 500, 600, 700, 800 nm	60%,70%,79%,75%,65% respectively
Output data rate	50 k pixels/s

Table 3 Secondary CCD Camera System Specifications

Type	Kodak KAF-1300 L
Format	1, 280 (H) pixels × 1, 024 (V) pixels
Pixel size	16 μm × 16 μm
Imaging area	20.5 mm (H) × 16.4 mm (V)
Dark current @25°C	<20 pA cm^{-2}
Readout noise	20 e^- rms
Full well signal	300.000 electrons
Dynamic range	>72 db
Quantum efficiency at room temperature	
λ = 500 nm	25%
λ = 600 nm	35%
λ = 700 nm	42%
λ = 800 nm	48%
Output data rate	20 MHz

The system was enhanced to allow remote operation of all the mechanical functions that were available on the unit. This involved focusing the telescope on the sky and the guider system then brought also to be in focus. This is done with a motorized and encoded eccentric cam arrangement. The position of the acquisition field is selected remotely by the operator and a motorized slide allows the guider field of view to be moved remotely and the position of the guider displayed. The main CCD has with it a 10-position filter wheel that is encoded to show the position of the wheel on the display. A set of photometric filters (UBVRI) is inserted in the wheel (Mackay 1995).

4.3 *Mirror Aluminizing Plant Upgrade*

In order to have high protection, efficiency and safe coating process for the new mirrors of the telescope, it was necessary to upgrade the existing aluminizing plant that was erected by EDWARDS, England in 1966. The plant was using the manual control and the obsolete analogue sensors and gauges of 1960s. As so and in 1998, the plant had an upgrading process that used BALZERS equipments. This included a new mechanical booster pump, a two stage oil sealed rotary pump, an exhaust mist filter, a higher throughput diffusion pump, a right angle baffle valve, a polycold, a backing valve, a roughing valve, a needle valve, a venting valve, a backing and roughing pipelines, and a new control cabinet. All valves used were pneumatically operated such that they all fail safe to the close positions in case of power failure, loss of water cooling supply, loss of compressed air, or pumping failure. The upgrade also included a very high efficient and intelligent cooling system, digital sensors and gauges and auto controlling all of the process by the up-to-date SIMENS Programmable Logic Control (PLC). The plant cell body, together with the filament electric power supply was the only part that did not have any change. Fig. 4 shows the upgraded part of the plant.

The new system has intelligent alarms for different expected problems related to cooling water supply cut off, reaching high temperature degrees, valves opening and closing pressures and emergency normal close of all the valves in the case of electricity failure. The system has been checked and tested thoroughly for pressure leakage, safety and for all aspects of failure operating conditions, and for the ultimate vacuum pressure that can be reached inside the plant cell. We usually start coating the mirrors by the aluminum material when the pressure reaches a value of 5×10^{-5} mbar at which case the pressure inside the cell may rise slightly. The new mirror of the telescope has been successfully coated by the upgraded system and various coatings for other test mirrors have been implemented by the system.

4.4 *Okayama-Kottamia Spectrograph (OKS)*

This spectrograph was originally designed for use on the 74-inch telescope of Okayama Astrophysical Observatory (OAO) in (Japan), with an image intensifier

Fig. 4 Upgraded Kottamia 74″ mirror aluminizing plant

and photographic plates. In the framework of cooperation between Japanese and Egyptian astronomers, it was suggested to present this spectrograph to be in use with the 74″ telescope of KAO. The essential modification was to replace the image tube system and photographic plate by a high performance CCD camera. This modification was designed by one of us (G. B. Ali) during his visit to OAO and was implemented in OAO. The spectrograph has been brought to KAO in Sep 2006. Two Japanese staff members of OAO, came to Egypt on Feb 2007 to help KAO staff in installing and testing of the instrument. After successful completion of their mission, the spectrograph was commissioned for operation at the f/18 Cassegrain focus of the 74″ telescope of KAO, see Fig. 5.

OKS is a low-to-moderate resolution spectrograph (700–1,200). It has a slit of 15 mm length corresponding to 90 arcsec on the sky. The slit is also of variable widths ranging from 0.005 to 3.0 mm corresponding to 0.03–18 arcsec on the sky. There is also a TV system to guide off the slit or in an offset mode. The instrument has a Hollow Cathode tube (Fe + Ne) used as a comparison lamp. The main specifications of the OKS are listed in Table 4.

For the 300 gr mm^{-1} grating, the spectral resolution at different order is:

		Resolution	
Order	Wavelength coverage over CCD	Spectral	Velocity
1	5,000 A	5.0 A	300 km sec^{-1}
2	2,500 A	2.5 A	150 km sec^{-1}
3	1,250 A	1.3 A	75 km sec^{-1}

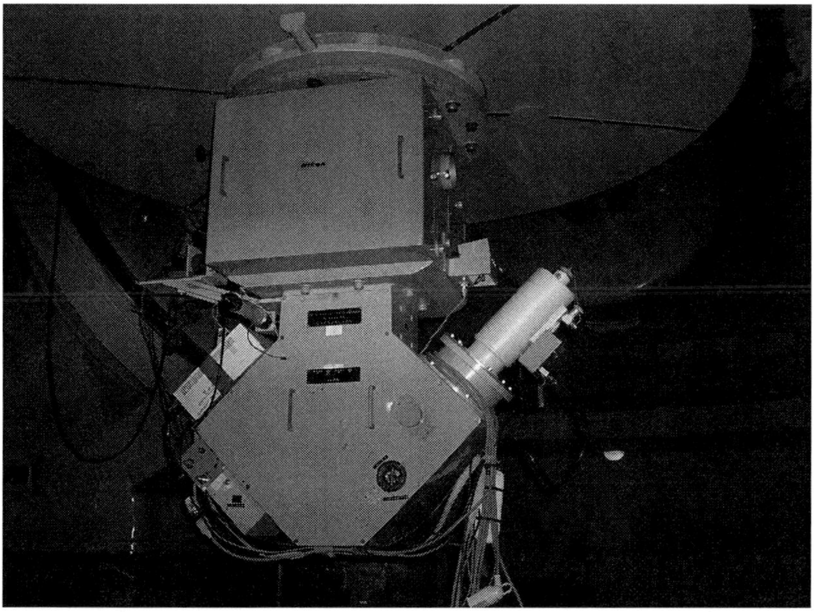

Fig. 5 OKS Spectrograph attached to the KAO's telescope

Table 4 Main Specifications of the OKS (Okayama-Kottamia Spectrograph)

Collimator	
Type	Inverse Cassegrain
Focal length	850 mm
Beam size	47.3 mm
Main mirror aperture	72 (70) mm
Secondary mirror aperture	24 (22) mm
F-ratio	F/12.1
Camera: Type	Solid Schmidt
Focal length	142.2 mm
F-ratio	F/2.5
Field of view	6 d
Slit: Width	5 μm∼3mm, one side open
Length	0.5∼15mm, both sides open
Comparison lamp	Hallow cathode tube (Fe + Ne)

A rough estimation of the exposure time for a 12th magnitude star is about ten minutes at first order. Four gratings are available to provide wavelength coverage from 4,000 to 8, 000 Å; their characteristics are given in Table 5.

Table 5 The available Gratings Characteristics

Catalog No.	n (lines/mm)	Blaze angle	Blaze λ (Å)	Dispersion at 1st order (Å/pixel)
35–53–11–170	300	4° 1	5,000	2.84
35–63–11–640	300	10° 2	12,000	2.84
35–53–11–560	600	22°	12,500	1.42
35–53–11–290	1,800	26° 4	5,000	0.47

Table 6 OKS CCD Camera Specifications

Type	EEV CCD 42–10 slow scan
Version	Back-illuminated (basic process), AIMO
Format	$2,048\,\text{pixels} \times 512\,\text{pixels}$
Pixel size	$13.5\,\mu\text{m} \times 13.5\,\mu\text{m}$
Grade	1
Imaging area	$27.6\,\text{mm} \times 6.9\,\text{mm}$
Dark current @20°C	76 e/pix/s
@ − 140°C	0.18 e/pix/hr
Readout noise @83.3 KHz	$3.4\,e^-/\text{pixel}$
Full well signal	96.000 electrons
Quantum efficiency at room temperature	
$\lambda = 350\,\text{nm}$	20.7%
$\lambda = 400\,\text{nm}$	48.3%
$\lambda = 500\,\text{nm}$	85.1%
$\lambda = 650\,\text{nm}$	85.9%
$\lambda = 900\,\text{nm}$	37.8%
Holding time	>12 h

4.4.1 Spectrograph CCD Camera System

The CCD camera of the spectrograph is an EEV CCD 42–10 grade 1, back illuminated installed in a downward looking high performance liquid nitrogen Dewar, its specifications are given in Table 6.

4.4.2 Test Observations

Some of the objects are observed during and after the test phase of the OKS, Fig. 6 gives some examples of the spectra for some of them as well as the spectra of the HCL comparison.

4.4.3 Suggested Scientific Programs

Considering the advantage of KAO, i.e. the latitude, the large number of clear nights and the availability of much machine, scientific observational programs could be achieved e.g., (1) Long term monitor programs of spectral variation of variable stars,

Fig. 6 Examples of test observation spectra of: (**a**) Saturn (Exp. T. is 1 min) (**b**) Alpha Leo (Exp. T. is 5 s) (**c**) Alpha Crab (Exp. T. is 5 s) and (**d**) the HCL comparison spectra of 2 s exposure time

AGNs, quasars, etc. (2) Follow up observation programs of transient phenomena such as novae, supernovae, those of catastrophic binaries, etc. (3) Spectral survey programs for objects of particular kind.

4.5 Telescope Control System Refurbishment

A refurbishment process for the whole control system of the telescope is currently in effect. The refurbishment process includes replacing the existing control cabinet with the up-to-date equipment enabling computer control of all aspects of the dome and telescope. The upgrade includes also replacing of the AC motors in RA and DEC axes with high resolution stepper motors, keeping the clutch and clamp motors in both axes working for the purpose of quick/slow conversion and also for the purpose of freeing the telescope for balance. Focus motors at Cassegrain and Newtonian foci will also be replaced with high resolution stepper motors. New high resolution absolute encoders at the RA, DEC, Cassegrain and Newtonian foci will be installed. The resulting accuracy of the telescope pointing is expected to be around 5 arcsec. The accuracy in adjusting the focus position is better than $\pm 5\,\mu$m. An RF hand paddle will be used to move the telescope and the dome to various positions. Encoders for the dome, upper shutter, lower shutter and viewing gantry will be installed to give their positions and to allow the dome to follow the telescope in its movements. A GPS receiver will be installed for precise timing and geographical coordinates. Two rack mount industrial computers will be installed with windows XP operating system. One computer is used for the telescope control and the other one for the new purchased science CCD which will be described in the following section. An autoguider with two stacked integrated filter wheels and a thermoelectrically cooled CCD will also be provided which has the facility to be installed in either the Cassegrain or the Newtonian foci. The software for the guider and filter wheels is fully integrated into the main telescope control system.

4.6 Kottamia New CCD Camera System

A new CCD camera has been purchased for the purpose of direct imaging either at the Cassegrain or at the Newtonian foci. The control software of this new CCD will be integrated in the telescope control software being recently implemented. This will give an ease of operation and control of both the telescope and the camera at the same time. A new industrial computer will be dedicated to the control of the camera and saving of the scientific images. The camera is provided by Retriever Technology, USA and has the specifications detailed in Table 7.

Table 7 Specifications of the new purchased Kottamia CCD

Type	EEV CCD 42–40
Version	Back-illuminated with BPBC (Basic Process Broadband Coating)
Format	2,048 by 2,048 pixels
Pixel size	13.5 μm by 13.5 μm
Grade	0
Dynamic range	30.000:1
A/D converter	16 bit
Imaging area	27.6 mm by 27.6 mm
Dark current @20°C	0.00029 e^-/pix/s
Readout noise @20 KHz	3.9 e^-/pixel
Full well signal	120.000 electrons
Spectral Range	200–1,060 nm
Quantum efficiency @ − 85°C	25.1% @ 350 nm
	65.5% @ 400 nm
	87.4% @ 500 nm
	80.3% @ 650 nm
	30.3% @ 900 nm
Cooling System:	LN to −100°C and better with temp control. Includes controller for stability within −/ + 0.1°C.
Holding time	12 h
On chip binning	1 × 1 till 8 × 8
PC interface	Fiber optic

Acknowledgment The authors would like to thank all colleagues who took part in the installation, developing and attending the tests made to various equipments of Kottamia telescope. In particular, we thank Mr El-Khamissy, Mr. Zeid, Mr. Elnaggar, and Eng. Imam. Many thanks is due to the two Japanese staff of Okayama astronomical observatory, Prof. Hiroshi Ohtani and Eng. Takafumi Okada for their great efforts and time they spent in Kottamia observatory and for their valuable explanations and discussions with the staff of Kottamia. Meanwhile we are very thankful to the director and the staff of Okayama observatory who helped in gifting and preparing the spectrograph for transfer to Kottamia Observatory.

References

Samaha, A. H., The Kottamia 74″ Telescope, Helwan Observatory Bull, 62 (1964)

Hassan, S. M., Upgrading the 1.9-m Kottamia Telescope, African Skies, 2, 16–17 (1998)

Issa, I. A., Gamal El Din, A. I., Astronomical Research Activities with the 74 inch Telescope at Kottamia Observatory, Astrophysics and Space Science, 118, 87–93 (1986)

Hassan, S. M., Kottamia Telescope Upgrading- from Developing Basic Space Science World Wide. A Decade of UN/ESA Workshops. In: Wamsteker, W., Albrecht, R. and Haubold, H. (eds.), pp. 237–240. Springer, Netherlands (2004)

Osman, A. M. I., Astronomy for Developing Countries, 24th meeting of the IAU, Special Session, Manchester, England, August (2000)

Mackay, C. D., Technical Manual for the Astronomical Imaging System Supplied to NRIAG by AstroCam Ltd., (1995)

Index

HELIOSPHERE
IHY
INTERNATIONAL HELIOPHYSICAL YEAR 2007

PLANETARY ENVIRONMENTS
IHY
INTERNATIONAL HELIOPHYSICAL YEAR 2007

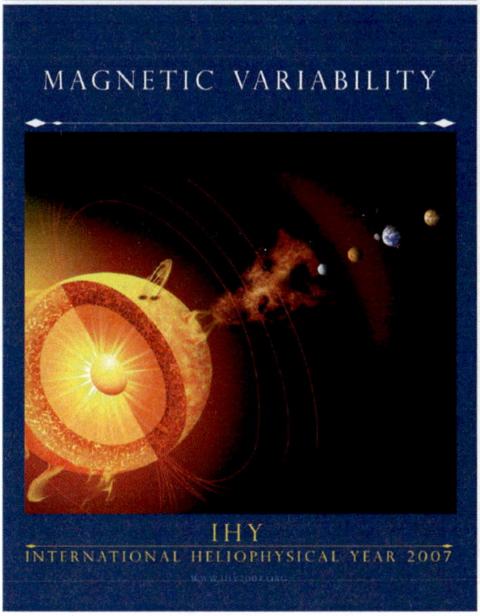
MAGNETIC VARIABILITY
IHY
INTERNATIONAL HELIOPHYSICAL YEAR 2007

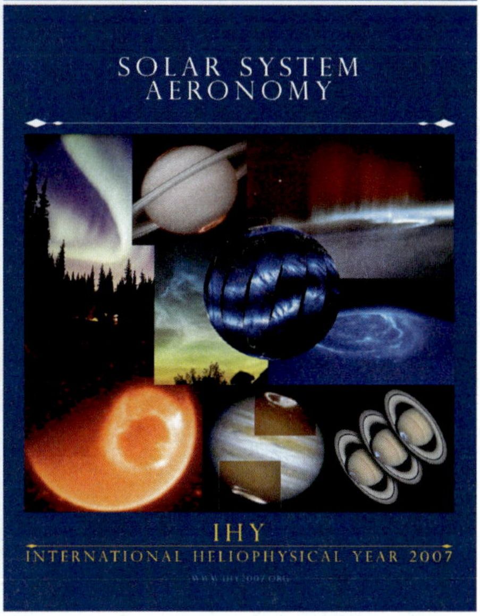
SOLAR SYSTEM AERONOMY
IHY
INTERNATIONAL HELIOPHYSICAL YEAR 2007